ASTON MARTIN

100 YEARS OF POWER, BEAUTY & SOUL

LENIN DEV

PHOENIX WINGS PRESS

COPYRIGHT

Published by Phoenix Wings Press LLC

For enquiries, please contact:

phoenixwingspress@gmail.com

FIRST EDITION

Table of Contents

Disclaimer

This book is intended for informational and educational purposes only. The author has made every effort to ensure the accuracy of the information presented; however, the automotive industry is continuously evolving, and details such as specifications, features, and corporate strategies may change over time.

The author and publisher disclaim any liability for any direct or indirect loss or damage resulting from the use of the information in this book. Readers are encouraged to conduct further research or consult official sources when making decisions related to Aston Martin vehicles.

This book is not affiliated with, endorsed by, or sponsored by Aston Martin Lagonda Global Holdings plc. The views and opinions expressed in this book are solely those of the author.

About the Author

Lenin Dev is a visionary automotive historian and bestselling author whose passion for speed, design, and engineering excellence drives every word he writes. With a flair for storytelling that fuses technical precision with cinematic elegance, Lenin captures the spirit and soul of legendary marques like Aston Martin, transforming their history into unforgettable narratives. His works have enthralled readers across more than 66 countries, with over one million copies in circulation worldwide, cementing his reputation as an authority on automotive artistry and motorsport legacy. From the racetrack to the studio, Lenin celebrates the harmony of power and beauty, revealing the heartbeat of machines that inspire generations.

Acknowledgments

Writing this book has been an extraordinary journey, and I am profoundly grateful to everyone who contributed to bringing it to life.

First and foremost, I wish to honor the visionaries behind Aston Martin—**Lionel Martin and Robert Bamford**—whose courage, passion, and dedication laid the foundation for one of the most iconic automotive brands in history. Their relentless pursuit of craftsmanship and performance continues to inspire generations of enthusiasts and engineers alike.

I am also indebted to the automotive historians, engineers, and enthusiasts whose meticulous research and tireless documentation preserved Aston Martin's rich legacy. Their insights and expertise were instrumental in shaping the narrative of this book.

Finally, to my readers—thank you for sharing in my passion for automotive excellence. It is my hope that this book deepens your appreciation for the artistry, innovation, and enduring **power, beauty & soul** that define Aston Martin.

1: THE GENESIS OF ASTON MARTIN

In the early years of the twentieth century, the world stood on the precipice of a technological revolution. Smoke-stacked factories hummed, steam trains raced across continents, and mankind's fascination with motion had reached fever pitch. The automobile was no longer a novelty—it was becoming a symbol of freedom, invention, and identity. Amid this awakening age of engineering, a name would soon be forged in metal, ambition, and speed: **Aston Martin**.

Even in its infancy, the brand carried an aura of distinction. It wasn't just another motor company born from the industrial frenzy; it was a dream made tangible by two visionaries—**Lionel Martin** and **Robert Bamford**. They weren't tycoons or aristocrats; they were craftsmen, tinkerers, and believers. Their creation wasn't merely a machine; it was an extension of human spirit and artistry, a fusion of performance and passion.

As Europe hurtled toward modernity, few could imagine that in a modest London workshop, a legacy was taking shape—one that would define British motoring excellence for over a century.

The story truly began in **1913**, when Lionel Martin and Robert Bamford joined forces to form a company called **Bamford & Martin Ltd.** The pair shared an unshakable belief that cars could be more than utilitarian objects—they could be experiences.

Lionel Martin, already known in motoring circles as an accomplished racer and meticulous engineer, brought with him an infectious enthusiasm for speed. He had raced at the famous **Aston Hill Climb** in Buckinghamshire, a place that tested both man and machine to their limits. The thrill of those races would later echo in every design that bore his name.

Robert Bamford, his partner, was the strategist—a businessman with a keen eye for quality and an understanding of what customers desired. Together, they embodied the perfect balance: Martin the visionary engineer, Bamford the grounded entrepreneur.

In a small workshop on Henniker Mews, London, they began sketching their first creation. The concept was simple yet profound: a car that combined **racing capability with elegant design**. They weren't building for the masses; they were crafting for the discerning few who demanded perfection.

When the first car rolled out of that workshop, its name was chosen not by committee or marketing analysis, but by heart and memory. Lionel Martin had conquered the Aston Hill, and thus, **"Aston Martin"** was born—a union of place, person, and passion.

But dreams, like engines, require fuel—and in 1913, that fuel was scarce. The world teetered on the edge of **World War I**, and the young company's ambitions were soon interrupted by global chaos. Both founders enlisted, halting production and redirecting their skills to the war effort.

The workshop fell silent, its unfinished chassis gathering dust. Yet, in that stillness, the spirit of Aston Martin endured. For Lionel Martin, every moment away from the track only strengthened his conviction. When peace finally returned in 1918, he came back not with despair, but with renewed purpose.

The post-war years were turbulent. Resources were limited, and the fledgling British car industry was fighting for survival. Yet Aston Martin refused to be ordinary. They could have built practical cars to secure quick profits—but that wasn't their calling. Instead, they continued to build **cars that stirred the soul**.

By 1921, Aston Martin unveiled its first production prototype—the **A3**. Its design was simple yet exquisite: light, balanced, and built for performance. It wasn't mass-produced; it was hand-crafted with precision, embodying the company's belief that **driving should be an art, not a routine**.

If engineering was Aston Martin's body, **racing was its heartbeat**. Lionel Martin's racing pedigree defined every car he touched. Competing wasn't a pastime for him—it was a crucible of progress. The racetrack was where metal met destiny.

From the **Aston Hill climbs** to the **Brooklands circuit**, Aston Martin's early racers began to make an impression. Their cars didn't always win, but they always inspired admiration. Drivers spoke of their responsiveness, their balance, and the unmistakable elegance even under stress.

When the company entered the **1922 French Grand Prix**, it wasn't just chasing victory—it was chasing validation. Competing against larger, better-funded teams, Aston Martin proved that craftsmanship and courage could rival capital. The car may not have finished first, but its performance signaled something important to the world: **Aston Martin had arrived**.

Each race was a laboratory. Every lap taught them about endurance, balance, and innovation. Failures became lessons; lessons became blueprints. Racing didn't just define Aston Martin's reputation—it refined its soul.

Yet, brilliance often walks hand in hand with struggle. The 1920s were an era of ambition shadowed by hardship. Economic instability, coupled with the financial strain of building specialized vehicles, put immense pressure on the young company.

By **1924**, Aston Martin faced bankruptcy for the first time. Many would have surrendered, sold the name, or walked away. But not Lionel Martin. He remained steadfast, guided by his belief that **greatness is not born of comfort but of conviction**.

The brand was rescued by a series of new investors, each of whom recognized its potential, even if they didn't fully grasp its soul. Through all the ownership changes and financial turbulence, Aston Martin's essence—its devotion to engineering artistry and high-performance elegance—remained untouched.

Martin once said, "A car should be beautiful even when it stands still." That philosophy transcended financial charts and business meetings. It became a creed. Even as the company restructured and regrouped, every model they built carried a sense of quiet majesty—metal shaped not just by machines, but by human hands guided by pride.

By the mid-1920s, Aston Martin's cars began to attract the attention of connoisseurs. Wealthy patrons saw in them something more than transportation—they saw identity. The blend of mechanical precision and understated elegance resonated deeply with the British sense of taste.

Every Aston Martin of that era was a **statement of individuality**. No two cars were truly identical, for each bore the subtle imprint of its makers. The brand's design philosophy became clear: restraint over extravagance, balance over excess. Where other luxury carmakers sought opulence, Aston Martin sought purity.

The chassis and engines were built for the track, but the finish was for the drawing room—a perfect equilibrium between aggression and grace. This duality became the hallmark of Aston Martin's DNA, a trait that continues even in modern times.

Racing continued to shape the company's evolution. By the late 1920s, Aston Martin's name was whispered with respect across European circuits. They weren't the fastest cars in raw power, but they possessed **balance**, **poise**, and **character**—qualities that defined the true gentleman's racer.

Every victory or loss was documented meticulously. Engineers returned from each competition with notes, sketches, and ideas. Adjustments were made to aerodynamics, suspension, and steering. The pursuit of improvement was relentless.

Through racing, Aston Martin learned that **speed is meaningless without control**. That ethos seeped into its road cars, making them not just fast, but deeply engaging to drive. Owners described the sensation as "symbiotic"—as if man and machine communicated in harmony.

As the 1930s dawned, the global landscape shifted again. Economic depression loomed, industries faltered, and the appetite for luxury waned. Yet even in uncertainty, Aston Martin continued to craft its masterpieces.

The company's survival was not due to chance—it was because of its unwavering **commitment to quality**. Customers trusted that every Aston Martin was built with integrity. Even as production numbers dwindled, the reputation grew stronger. It became the car of choice for the discerning few who valued distinction over display.

Through every hardship—financial collapse, global unrest, shifting ownership—Aston Martin endured because it stood for something eternal: **the celebration of motion, design, and courage**.

By the eve of World War II, Aston Martin had already achieved what many thought impossible. It had transformed from a two-man dream into a recognized emblem of British excellence. Its cars symbolized an ideal—that engineering and elegance could coexist without compromise.

The brand's very name had become a statement of purpose. "Aston" spoke of challenge and achievement; "Martin" embodied human dedication and artistry. Together, they formed a union that transcended mechanical creation—it was a philosophy.

Each nut, bolt, and curve carried meaning. To drive an Aston Martin was to participate in a legacy—one that began not in luxury boardrooms but in the sweat, imagination, and resilience of two dreamers who refused to settle.

As the 1930s closed and the specter of another war approached, Aston Martin once again faced the unknown. Production would halt, factories would quiet, and dreams would wait. But the flame never went out.

What began in 1913 as a daring experiment had evolved into a living legacy—a testament to human aspiration. The company's first quarter-century was not defined by profits or expansion, but by perseverance.

Aston Martin had learned the greatest lesson of all: that **a true marque is not built in a moment of triumph, but in the resolve to rise after every fall**.

Lionel Martin and Robert Bamford may not have lived to see the full extent of what their vision would become, but their spirit endures in every Aston Martin that graces the road today. Their belief—that cars could embody both emotion and excellence—became a guiding star for generations of engineers, designers, and drivers.

Theirs was a story not merely of business, but of artistry and conviction. Aston Martin was never about producing cars in quantity; it was about producing emotion in motion. That ideal—the pursuit of beauty, speed, and precision in perfect harmony—continues to define the brand's DNA more than a century later. And so, the legend was born—not from fortune, but from **fire, failure, and faith**.

When you hear the growl of an Aston Martin engine today—the measured symphony of pistons and exhaust—you're not just hearing machinery. You're hearing echoes of 1913, of Lionel Martin's hands on cold metal, of

Robert Bamford's belief that craftsmanship could transcend commerce.

Every Aston Martin that has ever existed owes its life to that single, daring dream. The company's birth was not just the start of a carmaker's journey; it was the genesis of a **philosophy of motion**—one that continues to inspire those who understand that true luxury lies not in abundance, but in authenticity.

Aston Martin was, and remains, a testament to **British engineering excellence**, to artistry under pressure, and to the eternal truth that **perfection is never finished—it is only pursued.**

The dawn of the 1940s brought both chaos and change. The world had once again plunged into the turmoil of war, and factories that once built dreams of speed were now devoted to survival. Like many great British marques, **Aston Martin's engines fell silent** as men and materials were diverted to the front lines. But silence, as history has shown, often conceals strength.

Inside those dormant workshops, the spirit of Aston Martin remained alive—a heartbeat beneath the dust, waiting to roar once more. The engineers and craftsmen who survived the war returned with a new sense of purpose. They had seen destruction; now they sought creation.

When peace finally came in 1945, Britain was a nation bruised but unbroken. Cities were rebuilding, industries were rekindling, and amidst the grey ashes of postwar austerity, **Aston Martin prepared for its rebirth**. What had begun as a vision of two men decades earlier was now poised to evolve into a global emblem of excellence.

In 1947, a turning point arrived. The company was purchased by **David Brown**, an industrialist whose initials—*DB*—would soon become legendary in the automotive world. Brown was not just a businessman; he was an enthusiast. He saw in Aston Martin something that transcended profit margins: a soul worth nurturing.

Under his leadership, Aston Martin entered a new golden age. Brown's engineering acumen and willingness to invest heavily in research gave the company a new foundation. He envisioned a line of vehicles that could **blend competition-bred performance with aristocratic luxury**, a rare combination that few could execute.

The first product of this new vision was the **DB1**, followed by the celebrated **DB2**—cars that carried both aesthetic grace and mechanical brilliance. These models redefined Aston Martin's identity. They were not merely fast; they were composed, elegant, and deliberate. To own one was to possess a piece of engineering artistry.

David Brown understood what Lionel Martin had always known: **motorsport was not just sport—it was survival**. Racing tested not only machinery but philosophy. Each victory refined the brand's character; each failure taught humility and progress.

In **1949**, Aston Martin returned to the track at the **24 Hours of Le Mans**. The cars were strong but unfinished; they showed promise, not dominance. Yet within a decade, Brown's relentless pursuit of excellence bore fruit.

The **DBR1**, unveiled in 1956, became a symbol of British racing brilliance. Sleek, balanced, and breathtakingly fast, it captured the imagination of enthusiasts worldwide. Its defining moment came in **1959**, when Aston Martin achieved its most glorious triumph: winning **Le Mans** outright, along with the **World Sportscar Championship.**

That year cemented the brand's racing credentials forever. The DBR1, driven by legends **Carroll Shelby** and **Roy Salvadori**, didn't just win—it conquered. The roar of its engine echoed across the world, signaling that Aston Martin had transcended its humble origins to become an icon of endurance and excellence.

Success on the racetrack fueled success on the road. The lessons learned from the DBR1 directly shaped the design and performance of Aston Martin's road cars. In 1958, the **DB4** debuted—a masterpiece that blended Italian styling by **Carrozzeria Touring of Milan** with British engineering finesse.

Its long bonnet, sweeping curves, and distinctive grille became instantly recognizable. It wasn't merely a car—it was sculpture in motion. Critics hailed it as one of the most beautiful cars ever made, a description that remains true to this day.

The DB4 also introduced Aston Martin to a new clientele—one that desired both **refinement and rebellion**. These were individuals who appreciated fine craftsmanship but demanded power and individuality. In the DB4, they found both.

Each successive model—**DB5**, **DB6**, and beyond—elevated the marque's prestige. The DB5, launched in 1963, would soon become immortalized through cinema, but even before the silver screen, it was already a marvel of design. Underneath its polished aluminum skin lay precision engineering and a symphony of speed that spoke to Aston Martin's relentless evolution.

In 1964, a cinematic phenomenon forever linked Aston Martin to cultural immortality. When **James Bond** drove the **DB5** in *Goldfinger*, an icon was born. It wasn't just product placement—it was perfect symbolism. Bond's character, suave yet deadly, mirrored Aston Martin's essence: elegance hiding formidable power. The DB5 became more than a car; it became **a statement of sophistication and danger**, embodying the fantasy of motion and mastery.

From that moment, the Aston Martin name was etched into global consciousness. People who had never driven one could recognize its silhouette. The brand became synonymous not just with luxury, but with *legend*. Every cinematic appearance that followed—each chase, explosion, and narrow escape—only deepened that association.

Yet, beyond Hollywood glamour, the DB5 represented something deeper. It captured the culmination of half a century of craftsmanship, a balance between beauty and purpose that no rival could replicate.

But success, like the open road, is never without obstacles. The late 1960s and 1970s brought economic turbulence that tested Aston Martin's endurance. Oil crises, rising costs, and shifting consumer habits challenged the company's financial foundation.

Ownership changed hands multiple times, each bringing new strategies and promises. Yet through every transition, Aston Martin refused to dilute its values. It continued to build cars that spoke of passion, not pragmatism. The production numbers were low, but exclusivity only heightened desirability.

During these years, models like the **V8 Vantage** emerged—a car that redefined power and aggression for the brand. Its muscular stance and thunderous performance made it one of the most formidable grand tourers of its era. The V8 era was raw, unfiltered, and unapologetically British—a declaration that Aston Martin would never compromise its identity for mass appeal.

While other manufacturers chased trends, Aston Martin remained loyal to its lineage. It was never about following—it was about leading quietly, with dignity and defiance.

The 1980s and 1990s marked a period of reinvention. The automotive world was changing rapidly—computer-aided design, aerodynamics, and new safety standards reshaped how cars were made. For Aston Martin, it was both a challenge and an opportunity.

Under the stewardship of new investors and later, partial ownership by the **Ford Motor Company**, Aston Martin gained access to resources and technology that allowed it to reimagine itself without losing its essence.

In 1994, the **DB7** debuted—a return to the elegance and emotional design that had defined the brand's golden years. Styled by **Ian Callum**, it embodied grace, balance, and timeless allure. The DB7 wasn't just a car—it was a bridge connecting past glory with future promise. It sold in numbers far greater than any previous Aston Martin, revitalizing the marque and proving that the world still craved authenticity in an age of automation.

Through every reinvention, Aston Martin held fast to one sacred principle: **each car must be built by hand**. In an industry that increasingly valued mass production and uniformity, Aston Martin remained defiantly artisanal.

In the factory at **Gaydon**, each car is assembled by master craftsmen who treat the process as an art form. The scent of leather, the gleam of aluminum, the rhythmic sound of tools—all blend into a quiet symphony of creation. Every stitch, every curve, every torque of a bolt carries intention.

To own an Aston Martin is to own a story—one written in precision and passion. It's not just transportation; it's transformation. That commitment to the human touch has ensured that every model, from the **Vanquish** to the **Vantage**, carries a sense of intimacy that no robot could replicate.

As the 21st century unfolded, Aston Martin reclaimed its rightful place among the most revered names in automotive history. The release of the **DB9** and the **DBS** reaffirmed its dominance in the luxury grand touring segment. Each model married technology with tradition, reminding the world that innovation need not erase heritage.

The DB9 in particular represented a new era of confidence. With its V12 heart and sculpted form, it was both futuristic and familiar—a car that respected the past while racing toward tomorrow.

Aston Martin's global expansion also grew stronger. From Beverly Hills to Dubai, its showrooms became sanctuaries for those who saw cars as poetry in motion. Yet, no matter how far its reach extended, the brand never lost its British soul.

True to its heritage, Aston Martin never abandoned its racing roots. The early 2000s saw the marque return to endurance racing with renewed vigor. The **DBR9**, built upon the DB9 chassis, roared back into Le Mans in 2005 and captured the **GT1 Class victory**.

This wasn't nostalgia—it was reaffirmation. The engineers and drivers who carried the Aston Martin badge back to Le Mans understood that they weren't just racing for trophies. They were racing for history, for legacy, for the pride of the winged emblem that symbolized flight through adversity.

That victory, and those that followed, proved that **Aston Martin's DNA remains unchanged**—refined by time, but never rewritten.

Modern Aston Martins—like the **DB11**, **Vantage**, and **DBS Superleggera**—carry the same spark that first ignited in 1913. Each model reflects a century's worth of evolution, not revolution. The curves are sleeker, the engines more advanced, yet the soul remains familiar.

Today's Aston Martin blends artistry with aerodynamics, technology with tradition. Its designers speak of "emotional purity," a concept that captures what the brand has always stood for—**cars that move the heart before they move the wheels.**

With ventures into electric innovation and limited hypercars like the **Valkyrie**, Aston Martin continues to redefine what British performance means in the modern age. Even as the automotive landscape evolves toward sustainability and digitalization, Aston Martin insists on maintaining its essence: beauty, balance, and bespoke craftsmanship.

To understand Aston Martin is to understand that it is more than a manufacturer—it is a philosophy. Its story is a meditation on **human ambition**, on the desire to create something timeless in a transient world.

From Lionel Martin's modest workshop to David Brown's racing triumphs, from the cinematic glamour of Bond to the electric future of Valkyrie, every chapter of Aston Martin's journey speaks to one enduring truth: **perfection is a pursuit, not a destination**.

Each car that leaves its factory is a quiet rebellion against mediocrity, a whisper of artistry in an era of automation. It reminds us that even in a world obsessed with speed, the real measure of excellence lies in soul.

Aston Martin's story began with two men and a dream. Over a century later, it continues with thousands of dreamers—engineers, designers, drivers, and enthusiasts—who carry that same fire in their hearts.

The winged badge has become more than a logo; it is a symbol of **enduring craftsmanship, courage, and class**. Every generation of Aston Martin bears the same promise: that when you turn the key, you're not just starting an engine—you're awakening history.

As the brand steers into its second century, it remains guided by the same five pillars that defined its beginning: **Passion, Precision, Perseverance, Performance, and Prestige**.

The Genesis that began on a quiet London street in 1913 has become an everlasting odyssey. The flame of Aston Martin still burns bright—not in factories or racetracks alone, but in the hearts of all who believe that driving can still be **an art form, a calling, and a dream made real**.

2: THE DB ERA BEGINS

The Second World War had left the world weary yet restless, hungry for progress, and determined to rebuild from the ashes of devastation. Across Europe, factories once dedicated to war machines were reawakening with a new purpose—to create vehicles of freedom, mobility, and inspiration. Among the British automakers rising from the rubble was Aston Martin, a company whose survival had often hung by a thread. What awaited it after the war, however, was a transformation so profound that it would forever define the marque's identity.

By 1946, Aston Martin was little more than a name—its production halted, its finances depleted, and its facilities worn. Yet, within the quiet hum of post-war England, destiny was stirring. It arrived in the form of an industrial magnate and visionary: **David Brown**. A man of mechanical genius and entrepreneurial courage, Brown would not only rescue Aston Martin from obscurity but forge it into a global icon of performance and elegance.

The post-war years were a time when the world sought beauty again—art, speed, and craftsmanship that spoke to the resilience of the human spirit. Aston Martin, reborn under Brown's stewardship, would deliver precisely that. With the introduction of the **DB series**, named after its new savior,

Aston Martin would redefine what a grand touring car could be: not merely transportation, but a statement—a fusion of power, prestige, and purpose.

David Brown was not just another businessman with an interest in fast cars. He was a mechanical craftsman with petrol in his veins. Born in Yorkshire in 1904, Brown inherited a thriving family engineering business, **David Brown & Sons**, known for building industrial gears and tractors. His technical background gave him a deep understanding of machinery, but it was his fascination with automobiles that would change the trajectory of his career—and Aston Martin's fate.

In 1947, Brown came across a small advertisement in *The Times* newspaper that read, "High-class motor business for sale." Intrigued, he investigated further and soon found himself standing before the modest facilities of **Aston Martin Lagonda Limited**. The company, though rich in racing history, was struggling financially. Yet Brown saw potential—an English jewel waiting to be polished. He acquired Aston Martin for just **£20,500**, a sum that would become one of the most legendary investments in automotive history.

But Brown's ambitions did not end there. In the same year, he also purchased **Lagonda**, another respected but struggling automaker, primarily to obtain its advanced **2.6-liter straight-six engine**, designed by the great engineer **W.O. Bentley** himself. With Aston Martin's chassis expertise and Lagonda's engine technology, Brown now possessed all the ingredients to build cars that could rival the best from Italy and Germany.

The world would soon come to know this fusion of engineering brilliance under two simple letters: **DB**.

In 1948, the first Aston Martin under David Brown's ownership made its debut: the **DB1**. Although only fifteen units were produced, the DB1 was far more than a commercial endeavor—it was a declaration of intent. The sleek, open-top roadster was designed to showcase what the new Aston Martin stood for: craftsmanship, refinement, and the promise of speed.

The car featured a 1.5-liter, four-cylinder engine and lightweight construction. Its performance, by modern standards, was modest—a top speed of around **90 mph**—but in the late 1940s, it represented the cutting edge of British sports car engineering. What truly set the DB1 apart was not its numbers, but its **character**. Every curve of its aluminum body was sculpted by hand, every bolt and rivet placed with meticulous care.

The DB1's debut at the **1948 London Motor Show** was a moment of quiet triumph. The crowds that gathered around its gleaming form saw not just a car, but the resurrection of a dream. For Aston Martin, which had nearly vanished into the fog of wartime bankruptcy, the DB1 was proof that passion could indeed rebuild what war had broken.

Even though the DB1's production run was limited, it set in motion the momentum that would propel the company into an era of unprecedented innovation and prestige.

David Brown understood something vital: if Aston Martin was to be respected, it had to prove itself on the racetrack. Motorsport was not merely about glory—it was about engineering evolution. Every victory, every mechanical failure, every grueling lap around Le Mans or Silverstone taught lessons that would refine the road cars to perfection.

Aston Martin's reentry into racing came swiftly. In 1949, the company entered the **24 Hours of Spa** with a modified DB1, testing its endurance and reliability. Though the team didn't secure victory, the event marked the beginning of a renewed racing heritage. Brown was not content with mediocrity—he wanted Aston Martin to stand shoulder to shoulder with the likes of Ferrari, Jaguar, and Maserati.

To achieve that, he needed more power, more speed, and more technological innovation. And so began the journey toward the **DB2**, a car that would catapult Aston Martin into the international spotlight.

Unveiled in **1950**, the **Aston Martin DB2** was the company's first truly modern grand tourer. It was the car that defined Aston Martin's DNA—sleek, sophisticated, and unrelentingly fast. Powered by the 2.6-liter **Lagonda straight-six engine**, the DB2 could achieve a top speed of **120 mph**, a figure that placed it among the elite performance cars of its era.

Its aerodynamic design was revolutionary for its time. With smooth contours and a flowing silhouette, the DB2 embodied an elegance that appealed to both racers and aristocrats. Inside, the cabin was trimmed in the finest Connolly leather, offering luxury that rivaled Rolls-Royce while maintaining the soul of a race car.

But the DB2 wasn't just about looks. It was built to dominate. In **1950**, Aston Martin entered three DB2s into the **24 Hours of Le Mans**, where they finished impressively high despite mechanical setbacks. The following year, a DB2 claimed victory in its class, announcing to the world that Aston Martin was not merely surviving—it was thriving.

The success of the DB2 ignited a new era of ambition. Aston Martin had proven it could compete at the highest level of endurance racing. Now, Brown and his engineers sought to build a machine that would conquer it all—the perfect blend of luxury and raw power.

If the DB2 was a grand tourer for the road, the **DB3** was a warrior for the track. Introduced in **1953**, it represented Aston Martin's uncompromising pursuit of racing supremacy. Designed specifically for competition, the DB3 featured a lightweight tubular frame, improved suspension, and a 2.9-liter engine producing over 140 horsepower—an impressive figure for the time.

Its purpose was singular: victory at **Le Mans**. Though the DB3 faced stiff competition from Jaguar's C-Type and Ferrari's 250 MM, it held its ground admirably, showcasing Aston Martin's relentless engineering spirit. It was not the ultimate Le Mans winner—that would come later—but the DB3's contribution to refining the company's racing expertise was invaluable.

The DB3's successor, the **DB3S**, arrived in **1953** as a lighter, faster, and more aerodynamic version. With its curvaceous bodywork and polished aluminum finish, it looked as if it had been sculpted by the wind itself. Driven by racing legends like **Stirling Moss** and **Peter Collins**, the DB3S brought Aston Martin tantalizingly close to championship glory, earning podium finishes and international acclaim.

Though Aston Martin would not achieve its ultimate racing triumph until the **DBR1** in 1959, these early models cemented its reputation as a builder of cars that were both thrilling to drive and breathtaking to behold.

One of the defining elements of the DB series was its aesthetic philosophy. In an age when most cars were built for function over form, Aston Martin treated design as an art form. Each model was a balance of aggression and grace—a reflection of the English gentleman racer archetype: refined, poised, yet always ready to unleash power.

Under David Brown's leadership, the company worked closely with some of the most talented designers and coachbuilders of the era. The influence of **Italian design houses**, particularly **Carrozzeria Touring of Milan**, would later define the DB4 and beyond, but even in the DB2 and DB3, the seeds of that elegance were already present.

Aston Martin's design ethos went beyond beauty; it was about proportion, symmetry, and identity. The long bonnet, short rear deck, and muscular haunches became hallmarks of the brand—instantly recognizable and eternally stylish.

Every Aston Martin built under David Brown carried a distinct aura of **British craftsmanship**. The cars were assembled by hand in **Feltham**, later moving to **Newport Pagnell**, where artisans, not machines, shaped metal and stitched leather with an attention to detail that bordered on obsession.

To own an Aston Martin during the 1950s was to own a piece of automotive art. The clientele ranged from aristocrats and actors to racing enthusiasts who valued refinement as much as performance. The DB cars were not ostentatious—they were quietly commanding, exuding confidence rather than shouting for attention.

This understated elegance became a defining trait of Aston Martin's identity, setting it apart from the flashier Italian marques. Where Ferrari celebrated flamboyance, Aston Martin embodied **discretion and distinction**—the perfect car for those who wanted to be noticed without trying to be seen.

By the mid-1950s, Aston Martin had found its rhythm. The DB2 had cemented its credibility, the DB3 and DB3S had sharpened its competitive edge, and its name was whispered in the same breath as Europe's most elite manufacturers. Yet, David Brown was not satisfied. He envisioned a car that would transcend the boundaries between road and race—a car so beautiful, so fast, and so perfectly balanced that it would define the very essence of Aston Martin.

That car would arrive in **1958**. Its name: **DB4**. But before that momentous leap, Aston Martin had to consolidate its hard-earned lessons, refine its engineering capabilities, and prepare to take on the world not just in racing circuits but in the hearts of drivers everywhere. The DB era had begun in earnest, and the world was watching.

By the late 1950s, the world of automobiles had entered a golden age. The war was long behind, prosperity was blooming, and style had become a form of power. It was in this atmosphere of optimism that Aston Martin unveiled the **DB4** in **1958**—a machine that didn't just mark a new chapter for the company but an entirely new identity.

The DB4 was revolutionary. It was the first Aston Martin to be **entirely built at the new Newport Pagnell factory**, signaling both progress and permanence. But what truly distinguished the DB4 was its **design partnership** with **Carrozzeria Touring of Milan**, the famed Italian coachbuilder known for its lightweight "Superleggera" construction method. The result was a car that merged British engineering integrity with Italian artistry—a masterstroke that redefined grand touring.

The DB4 was not merely beautiful; it was breathtaking. Its proportions were perfect—long, sculpted hood, a taut rear, and a grille that seemed to smirk with quiet superiority. Beneath its bonnet lay a 3.7-liter straight-six engine, capable of propelling the car to **153 mph**, a figure that made jaws drop across Europe. It was one of the fastest cars in the world.

Each detail was considered with obsessive precision: hand-polished aluminum panels, sumptuous Connolly leather interiors, and gauges that glowed softly against walnut veneers. The DB4 wasn't just a car; it was a statement of national pride—**Britain's answer to Ferrari and Maserati**, yet distinctly English in temperament.

The DB4's success catapulted Aston Martin into the pantheon of elite manufacturers. But for David Brown, this was not the destination—it was a foundation. He wanted more. He wanted a car that could embody everything Aston Martin had become—its performance, its elegance, its mystique—and project it to the entire world. That car would become one of the most iconic automobiles in history.

In **1963**, Aston Martin unveiled what many consider the most famous car ever built: the **DB5**. From the moment it appeared, it was destined for greatness. Everything about it—the design, the engineering, even its name—carried an aura of destiny.

The DB5 was a natural evolution of the DB4 but elevated in every dimension. It housed a **4.0-liter engine** producing

282 horsepower, giving it blistering performance for its time. The car could sprint from 0 to 60 mph in **8 seconds** and reach a top speed of **145 mph**—numbers that placed it among the world's fastest production cars.

But the DB5's real power wasn't measured in acceleration; it was measured in allure. Its polished chrome details, refined contours, and sense of timeless proportion gave it a kind of cinematic quality. It was elegance incarnate—a car equally at home roaring down Le Mans straightaways or gliding through the streets of London's Mayfair.

And then, in **1964**, the DB5's fate was sealed forever when it became **James Bond's car of choice** in the film *Goldfinger*.

The collaboration between Aston Martin and the 007 franchise was nothing short of magical. In *Goldfinger*, Sean Connery's Bond piloted the silver birch DB5 equipped with gadgets that thrilled audiences—machine guns, revolving license plates, radar screens, and the legendary ejector seat. The car didn't just steal scenes; it stole hearts.

The DB5 became more than a luxury sports car—it became a **symbol of cool**. Every man wanted to drive one, every woman wanted to be seen in one, and every carmaker envied it. The association with Bond transformed Aston Martin from a niche marque into a cultural phenomenon.

The car's waiting list grew longer than ever. Celebrities, industrialists, and royals clamored to own what was now the most desirable automobile on the planet. Yet, beneath the glitz, the DB5 remained a pure driver's machine. Its balance, responsiveness, and growling exhaust note ensured that the Bond glamour never overshadowed the soul of the car—a true Aston Martin, through and through.

What made the DB5 truly exceptional was the way it merged **engineering excellence** with **artistic craftsmanship**. While Italian exotics of the time dazzled with flamboyant speed, the Aston Martin offered something subtler—a quiet confidence born from precision and restraint.

Each DB5 was hand-built at Newport Pagnell by a small team of craftsmen. The car's aluminum body panels were beaten into shape by hand, fitted over a lightweight tubular frame, and finished with coats of deep, lustrous paint that took weeks to perfect. Inside, the cabin felt like a gentleman's club—supple leather, wool carpeting, polished wood, and chrome switches that clicked with satisfying mechanical precision.

No two DB5s were exactly alike. Each bore the fingerprints, literally and figuratively, of the artisans who built it. This level of care made the car feel alive—more than metal and gears, it had character.

Performance, too, was uncompromised. The DB5's 4.0-liter engine delivered torque in abundance, allowing effortless acceleration even at high speeds. Its ride was supple, its handling confident, and its brakes—disc brakes all around, still a novelty at the time—provided remarkable stopping power. This was the essence of Aston Martin: beauty married to performance in perfect balance.

When the **DB6** was introduced in **1965**, it marked the maturation of the DB lineage. The new model retained the elegance of its predecessors but added refinement and practicality. Its longer wheelbase improved cabin space, allowing the DB6 to serve as both a high-speed grand tourer and a luxurious daily driver.

The DB6's most distinguishing feature was its **Kamm tail**, a truncated rear design inspired by aerodynamics research from the world of motorsport. This gave the car greater stability at high speeds and became one of the most recognizable silhouettes in automotive history.

Under the hood, the DB6 retained the mighty 4.0-liter engine from the DB5 but offered an optional **Vantage specification** that boosted output to nearly **325 horsepower**. For those who desired even more exclusivity, the **DB6 Volante**, a convertible version, provided open-air motoring at its most glamorous.

While the DB5 had captured the imagination of moviegoers, the DB6 captured the hearts of drivers. It was a car for connoisseurs—mature, poised, and perfectly balanced. Many distinguished owners, from **Paul McCartney** to **Prince Charles**, fell under its spell.

The DB6 was not as radical as its predecessors, but it was evolutionary perfection—the culmination of everything Aston Martin had learned during its golden decade.

Even as the DB cars captivated the world's elite, Aston Martin continued its relentless pursuit of racing excellence. The **DBR1**, a purpose-built racing machine derived from the DB lineage, achieved what David Brown had long dreamed of: **victory at the 1959 24 Hours of Le Mans**. Driven by **Carroll Shelby** and **Roy Salvadori**, the DBR1 conquered the world's toughest endurance race, cementing Aston Martin's reputation as both a luxury manufacturer and a racing powerhouse.

This triumph reverberated through every model that followed. The DB4, DB5, and DB6 all bore the spirit of that victory—precision engineering honed by competition, refined by artistry. Racing was never just about trophies for Aston Martin; it was the company's proving ground, its forge of innovation.

Technological advancements from the racetrack found their way into road cars—better suspension systems, improved cooling, lighter materials, and more efficient engines. The lessons of Le Mans became the DNA of every DB built thereafter.

To drive a DB series Aston Martin in the 1960s was to experience something transcendent. The start of the engine was a performance in itself—a deep, throaty rumble that rose into a sonorous crescendo. The gearshift, mechanical yet precise, demanded engagement. The steering, heavy at low speeds but alive with feedback on the open road, invited mastery rather than mere operation.

This was no ordinary car. It was an experience—an orchestra of sound, motion, and emotion. Every mile in a DB was a reminder that engineering could be poetic, that machinery could stir the soul.

In an age when motoring was transitioning from art to industry, Aston Martin remained proudly artisanal. It built not for the masses but for the passionate few who understood the difference between driving and being driven.

By the end of the 1960s, the DB series had become more than a product—it had become an icon. Its presence in film, its association with the Bond mystique, and its appearance in the garages of rock stars and aristocrats elevated Aston Martin into a realm of cultural immortality.

The DB5, especially, transcended generations. It reappeared in later Bond films—*GoldenEye*, *Skyfall*,

Spectre—each time reminding audiences of a bygone era when cars were not just machines but symbols of identity and aspiration.

In art, literature, and even fashion, the DB's aesthetic became shorthand for **refined rebellion**—the union of sophistication and danger. The car represented what every man secretly desired to be: stylish, fearless, and irresistibly British.

The DB lineage did not end with the DB6. Its spirit carried forward through decades, reborn in models like the **DB7**, **DB9**, **DB11**, and the **modern DBS Superleggera**—each one an homage to the vision David Brown ignited.

The DB7, launched in 1994, resurrected Aston Martin during another period of hardship, just as Brown had done nearly half a century earlier. The DB9 of the 2000s redefined grand touring for the modern era—sleek, powerful, and luxurious. And the DB11, launched in 2016, stands as the 21st-century embodiment of that same philosophy: performance with elegance, power with grace.

Through every reinvention, the **DB badge** has remained sacred—a promise of excellence, a whisper of heritage. It connects generations of dreamers, collectors, and drivers who understand that an Aston Martin is more than transport—it's an expression of art, engineering, and emotion.

Even after David Brown sold Aston Martin in the 1970s, his initials lived on, immortalized in metal and myth. The "DB" wasn't just a model designation—it was a legacy, a philosophy, a signature etched into history.

Brown's vision—to build cars that embodied both **performance and poise**—became the company's compass. He proved that British engineering could stand toe-to-toe with

Italy's passion and Germany's precision, not by copying them but by being distinctly, unapologetically English.

Today, every Aston Martin that bears the DB monogram carries a part of his spirit—ambitious, exacting, and elegant. The initials "DB" have become inseparable from the brand itself, representing not just a man but an idea: that the perfect car must be as thrilling to behold as it is to drive.

The **DB era** defined Aston Martin's soul. From the tentative rebirth of the DB1 to the world-dominating glamour of the DB5, and the graceful refinement of the DB6, each car told a chapter of resilience, craftsmanship, and courage. Together, they forged a lineage that transcended the confines of the automobile industry to become a cultural and artistic statement.

More than seventy years after the first DB1 rolled onto the streets, the DB name continues to inspire reverence. Collectors pay fortunes for pristine examples, while designers and engineers still look to those mid-century masterpieces for guidance. They remain benchmarks of balance—machines that blend mechanical precision with human emotion.

The DBs were not just cars; they were **ambassadors of British excellence**, symbols of a nation's ingenuity reborn from the trials of war. Their story reminds us that greatness is not built in a day—it's crafted, refined, and passed down, one masterpiece at a time.

As the 1960s drew to a close, the world was changing fast. New materials, new technologies, and new social revolutions were reshaping everything—from art to industry. Yet, amid this transformation, Aston Martin's DB series stood as a reminder of timeless values: craftsmanship, passion, and integrity.

Each DB, from the earliest prototype to the latest supercar, tells the same story—a story of artistry and ambition bound by steel and soul. To drive one is to touch history; to hear its engine roar is to feel the heartbeat of a century.

The **DB era** was not just the beginning of modern Aston Martin—it was the defining heartbeat that continues to pulse through every car that bears its wings. And as long as there are roads, dreams, and the desire to chase perfection, the spirit of the DB will endure.

3: A RACING PEDIGREE

In the great symphony of automotive history, few names have resonated with such enduring rhythm as **Aston Martin**. It is a name that conjures visions of power surging down long straights, of gloved hands gripping a trembling wheel, of engines roaring not merely for glory but for survival. While the marque's cars have long adorned the driveways of the world's elite, their soul was forged not in luxury showrooms but on the punishing tarmac of the racetrack. From its earliest days, Aston Martin was born not as a carmaker alone, but as a racer — a dream molded by competition, risk, and the relentless pursuit of perfection.

The story of Aston Martin's racing heritage is one of resilience, rebirth, and triumph against overwhelming odds. Every era, from its tentative beginnings in the 1920s to its golden victories in the 1950s, tells a tale of ambition that defied circumstance. Where others sought mere success, Aston Martin sought something rarer — immortality through performance.

When Lionel Martin and Robert Bamford first built their modest racing machine in 1913, they could hardly have imagined the legend they were setting in motion. The name "Aston Martin" itself was born from **the Aston Hill Climb**, a grueling test of both man and machine near Buckinghamshire.

For Lionel Martin, victory at that hill was not just a pastime — it was proof of what precision, skill, and courage could achieve. From that moment onward, the company's ethos would forever intertwine with the thrill of racing.

In those early decades, the racecourses of Britain and Europe became laboratories of innovation. Each bend, each straight, each grueling endurance test pushed the fledgling engineers to refine, rebuild, and rethink. Racing wasn't a marketing exercise; it was survival. The early Aston Martins were small, hand-built machines, powered as much by passion as by petrol. Yet even then, they displayed the unmistakable hallmarks of what would become the brand's defining character: elegance fused with ferocity.

The interwar years were turbulent for the company, but motorsport remained its heartbeat. Even when resources ran thin, Aston Martin cars were entered into competitions, driven by enthusiasts and privateers who believed in their magic. Every race became an act of faith — a declaration that Aston Martin belonged among the greats.

Among all the arenas of motorsport, **Le Mans** stands as the ultimate trial. The **24 Hours of Le Mans**, with its combination of speed, stamina, and suffering, is not merely a race — it is a test of willpower and engineering unlike any other. Aston Martin's fascination with the event began as early as 1923, when the marque joined the grid for the first time. The outcome was modest, but the experience unforgettable.

For Aston Martin, Le Mans became both a proving ground and a spiritual home. The brand's engineers learned that victory was not won in a single lap, but in surviving the night — in crafting machines that could endure relentless mechanical and human punishment. Through the 1930s, Aston Martin's participation grew more serious, though financial troubles often interrupted their campaigns. Yet even in defeat, the company gathered knowledge, courage, and conviction. Every setback was a blueprint for future success.

Then came **the 1950s**, a decade that would transform Aston Martin forever. Under the visionary leadership of **David Brown**, the company re-entered the world of endurance racing with newfound determination. Brown understood that Le Mans was more than a race — it was a stage where legends were made. To dominate there would mean to elevate Aston Martin into a pantheon reserved for the likes of Ferrari and Jaguar.

The early attempts were brave but fraught with challenges. Engines overheated, gearboxes failed, and mechanical gremlins haunted the pits. Yet through each trial, the engineers refined their craft, and the cars grew stronger. By 1958, the DBR1 — a sleek, sculpted masterpiece powered by a straight-six engine — stood ready to make history.

The **1959 24 Hours of Le Mans** was a crucible of fire and glory — a race that would forever define Aston Martin's place in motorsport. Behind the wheel of the green-and-yellow DBR1s were two duos destined for greatness: **Carroll Shelby and Roy Salvadori** in one car, and **Maurice Trintignant and Paul Frère** in another. The competition was fierce. Ferrari, Porsche, and Jaguar had arrived with machines of formidable power. Yet Aston Martin's team carried something deeper — a quiet, unyielding belief in their car.

From the very start, Le Mans tested both men and machine to their limits. Day turned to night, and the relentless roar of engines merged into a hypnotic symphony of endurance.

Shelby, already a legend in the making, drove with reckless precision, pushing the DBR1 to its absolute limits while Salvadori held steady under immense pressure. Lap after lap, the team clawed its way toward the front, capitalizing on rival misfortunes and their own flawless execution.

By dawn, exhaustion hung heavy over the circuit, but the DBR1 remained unbreakable. Its engine sang with mechanical perfection, its body cutting through the early morning mist like a dart of emerald fire. When the checkered flag finally fell, it was the **Aston Martin DBR1** that crossed the line first — securing **Aston Martin's first and only overall Le Mans victory**. Salvadori and Shelby had not just won a race; they had etched their names — and the company's — into eternity.

That victory was more than a trophy. It was vindication. It was the culmination of decades of dreams, failures, and relentless evolution. It proved that Aston Martin could not only build cars of beauty but cars that could conquer the most brutal racing stage on Earth.

The **DBR1** remains, to this day, one of the most revered race cars ever built. It was not merely fast — it was alive, breathing elegance and aggression in perfect harmony. Beneath its sleek aluminum skin lay a 2.9-liter straight-six engine that produced around 250 horsepower — modest by today's standards, but revolutionary for its time. The car's lightweight tubular frame and aerodynamic profile made it as graceful on corners as it was fierce on straights.

What made the DBR1 extraordinary wasn't just its technical prowess but its soul. Every component had been meticulously designed to achieve balance — the perfect marriage of speed and stability. When drivers like Shelby spoke of the DBR1, they didn't describe it as a machine but as a partner — one that demanded respect and rewarded bravery.

Aston Martin's engineering team, led by **Ted Cutting**, approached the DBR1 as a living experiment in the laws of motion. The chassis design was light yet rigid, the suspension tuned for endurance rather than sheer velocity. The car's design philosophy would influence every Aston Martin that followed, from the DB4 to the modern-day Vantage. The DBR1 was not only a racer — it was the blueprint of Aston Martin's future.

Following its triumph at Le Mans, Aston Martin found itself at the summit of global motorsport. For a brief, shining moment, the world's eyes were fixed on a small British company that had defied giants. Yet, as is so often the case in motorsport, glory came at a price. Success on the track had drained enormous financial and technical resources. David Brown, pragmatic as ever, began to question the sustainability of such costly endeavors.

By 1960, Aston Martin withdrew from factory-backed racing. The team's departure marked the end of an era — one defined by daring, brilliance, and unyielding ambition. But though the company stepped back from the circuit, its racing spirit did not fade. Privateers continued to race Aston Martins across Europe, carrying forward the torch of excellence that the DBR1 had ignited.

Even in absence, the echoes of victory resonated through the halls of Aston Martin's workshops. Engineers continued to apply the lessons of endurance — building cars that were not merely beautiful but capable of withstanding the stresses of real-world performance. The company's road cars began to inherit the genes of the racetrack, blending comfort with competitive edge. The DB lineage, from the DB4 to the DB6, carried within it the DNA of Le Mans.

Aston Martin's **1959 Le Mans triumph** remains a defining chapter in the brand's mythology. It wasn't just about the victory — it was about what it represented. It told the world that Aston Martin, a company of craftsmen and dreamers, could stand shoulder to shoulder with industrial titans. It showcased a uniquely British blend of elegance and aggression, proving that performance need not come at the cost of beauty.

The ripple effects of that victory extended far beyond motorsport. It solidified the company's reputation as a creator of "gentlemen's racers" — cars that offered the thrill of competition and the comfort of class. The very notion of what a luxury sports car could be was redefined. No longer were racing and refinement opposing ideals; in Aston Martin, they coexisted.

To this day, the spirit of that era continues to pulse through the brand's veins. Every modern Aston Martin — from the DB11 to the Valkyrie — carries a trace of the DBR1's soul. Every rev, every curve, every shimmer of its polished surface is a silent homage to that long, victorious night at Le Mans. It was there, under the flickering glow of pit lights and the deafening echo of engines, that Aston Martin's heart began to beat — not as a manufacturer, but as a legend.

The 21st century marked a defining rebirth for Aston Martin's racing program. After decades of sporadic participation, the company returned to endurance racing with renewed vigor and a clear mission: to reclaim its place among the elite. In 2005, the formation of **Aston Martin Racing**

(AMR) transformed the brand's motorsport ambitions into a fully realized, professional enterprise.

This wasn't just a comeback; it was a statement of intent. The racing division was established in collaboration with Prodrive, one of the most successful motorsport engineering firms in the world. Together, they set out to build machines that could compete with—and beat—the giants of endurance racing. The result was the **DBR9**, a car that combined the timeless elegance of Aston Martin's design language with cutting-edge racing technology.

The DBR9 made its debut at the 2005 Sebring 12 Hours, immediately capturing attention. With its muscular stance, carbon fiber bodywork, and thunderous V12 engine, the car exuded power and sophistication in equal measure. More importantly, it was fast—blisteringly fast. The DBR9 soon proved itself on the global stage, securing class victories at both **Le Mans** and the **FIA GT Championship**, reigniting the flame that had burned brightly since the DBR1's triumph in 1959.

The next decade saw Aston Martin Racing cement its status as one of the leading names in endurance motorsport. Through persistence, innovation, and a relentless pursuit of perfection, the team became a symbol of resilience and engineering brilliance.

The **V8 Vantage GTE**, introduced in the early 2010s, became the face of Aston Martin's modern racing success. With its aerodynamic finesse and near-perfect weight distribution, the car embodied the ideal balance between power and precision. Competing in the **FIA World Endurance Championship (WEC)**, it delivered unforgettable moments of victory—most notably at **Le Mans 2017**, when the Vantage GTE clinched a dramatic class win in the closing minutes of the race.

The emotional intensity of that victory resonated far beyond the racetrack. It wasn't merely about speed; it was about endurance, teamwork, and the unyielding human spirit. The win was celebrated by fans and industry insiders alike as a triumph of both heart and engineering—a testament to Aston Martin's commitment to excellence.

After years of speculation, Aston Martin's return to **Formula 1** in **2021** marked the dawn of a bold new era. It was more than a racing initiative—it was a declaration of ambition. Partnering with the rebranded **Aston Martin Aramco Cognizant Formula One Team**, the company reignited its connection to the most technologically advanced and globally celebrated motorsport in existence.

The re-entry into Formula 1 represented Aston Martin's evolution from a handcrafted luxury marque into a forward-thinking performance powerhouse. The team's arrival on the grid was accompanied by striking British Racing Green livery—a nod to the brand's heritage and a symbol of its renewed vigor.

The move also aligned perfectly with the brand's strategy to transition into the hybrid and electric future of performance. Formula 1, with its advanced energy recovery systems and aerodynamic technologies, offered the ideal proving ground. Each race became a laboratory for innovation—testing new materials, refining software-driven performance systems, and exploring hybrid power solutions that would later influence Aston Martin's road cars.

One of the most fascinating aspects of Aston Martin's racing heritage is how seamlessly innovation flows from the racetrack to the road. The company's motorsport involvement isn't a separate endeavor; it's a developmental ecosystem that fuels progress.

Every high-speed lap, every grueling endurance race, and every pit-lane strategy session contributes to the evolution of the brand's production vehicles. Technologies that originated in racing—such as **carbon ceramic braking systems**, **aerodynamic diffusers**, and **adaptive suspension control**—have become integral to modern Aston Martin models.

Take the **Valkyrie**, for example—Aston Martin's hypercar born from Formula 1 DNA. Co-developed with Red Bull Advanced Technologies, the Valkyrie is more than just a car; it's a road-legal manifestation of pure racing engineering. Its naturally aspirated V12 engine, paired with hybrid assistance, delivers an experience that blurs the line between racetrack and reality.

In every way, the Valkyrie represents the philosophy that defines Aston Martin: *to build cars that stir the soul as much as they challenge the limits of performance.*

Motorsport has also deeply influenced Aston Martin's aesthetic language. The company's design ethos—fluid lines, aerodynamic grace, and sculpted power—has been refined by decades of racing experience. Each contour of an Aston Martin isn't merely artistic; it's functional, born from the wind tunnels and racetrack data.

Designers at Aston Martin work hand in hand with engineers to ensure that every detail serves both beauty and performance. The result is a fleet of vehicles that combine elegance with aggression—cars like the **DBS Superleggera**, **Vantage F1 Edition**, and **DB12**, which embody both luxury and latent ferocity.

This harmonious blend of design and technology is what distinguishes Aston Martin from its rivals. Where others may focus purely on power figures, Aston Martin pursues an emotional connection—a symphony of sound, speed, and style crafted for the driver who seeks more than just velocity.

What truly makes Aston Martin's racing pedigree extraordinary is its translation into the brand's everyday driving experience. Even in their most refined grand tourers, the essence of the racetrack is palpable. The roar of the engine, the precise feedback from the steering wheel, and the balance of the chassis all speak to decades of competition-honed expertise.

Every Aston Martin is imbued with the DNA of endurance and the adrenaline of speed. Whether it's the track-focused **Vantage GT4**, the luxurious **DB11**, or the commanding **DBX707**, each car carries a whisper of Le Mans and a heartbeat of Formula 1.

This is not by accident. Aston Martin's engineers deliberately calibrate performance characteristics inspired by racing—sharp throttle response, dynamic cornering balance, and the symphony of mechanical harmony that defines true driving pleasure. The result is a road car that doesn't just move; it *communicates*—a visceral dialogue between man, machine, and motion.

From **Carroll Shelby and Roy Salvadori's** heroic 1959 victory to the modern-day drivers conquering endurance circuits, Aston Martin's legacy is written in sweat, speed, and triumph. The men and women behind the wheel, in the pits, and at the design tables all share one common purpose—to uphold the spirit of a marque that has always strived to be both beautiful and brutally fast.

In a world where racing often defines the soul of a brand, Aston Martin's heritage stands among the greatest. Each victory is not merely a statistic; it's a chapter in a living legend—a story of relentless pursuit, audacious dreams, and British brilliance.

As Aston Martin accelerates into the future, the horizon gleams with promise. The next generation of racing will embrace electrification, sustainable materials, and intelligent systems that blend emotion with efficiency. Aston Martin is already preparing for this new era with its **Valhalla** hybrid supercar and its continued Formula 1 development program.

Yet, even as the world changes, one truth remains constant: the brand's devotion to racing excellence. Motorsport will continue to serve as the crucible in which Aston Martin's next innovations are forged. Whether it's a thunderous V12 or a silent electric surge, the goal remains the same—to thrill, to inspire, and to lead.

The racetrack is where Aston Martin's heart beats loudest. And from that heartbeat, the company draws the rhythm that powers every car it builds—a symphony of passion, power, and performance that defines one of the most illustrious names in automotive history.

4: THE JAMES BOND EFFECT

When the silver screen and the open road collided in the mid-1960s, something extraordinary was born—a partnership that would forever alter the landscape of both cinema and automotive culture. Aston Martin, the quintessential British luxury performance marque, found itself intertwined with one of fiction's most magnetic characters: **James Bond, Agent 007**. Together, they forged a legend that would outlast generations, spanning decades of cultural transformation, political upheaval, and technological revolution.

The bond—both literal and symbolic—between Aston Martin and 007 was far more than product placement. It was a meeting of two icons that shared a philosophy: elegance without arrogance, power without pretense, and danger cloaked in sophistication. When *Goldfinger* premiered in 1964, the cinematic world watched in awe as Sean Connery's Bond introduced the Aston Martin DB5 to the global stage. It wasn't just another car—it was a character.

For Aston Martin, this was destiny disguised as opportunity. The DB5 embodied everything that defined Bond's character—suave sophistication, impeccable taste, and lethal potential beneath a polished surface. It was as if Ian Fleming had designed the car himself to match the persona he had crafted in his novels. The public's reaction was instantaneous. Within months, Aston Martin had gone from a boutique British brand known to racing enthusiasts and aristocrats to an international symbol of style and power.

Bond didn't just drive the DB5—he *commanded* it. Each turn of the wheel, each quiet ignition of its inline-six heart, spoke to a man who mastered both technology and chaos. The DB5, glinting in the cinematic light of *Goldfinger*, represented a moment of pure alchemy: film and machinery merging to create myth.

The choice of the DB5 was not a mere coincidence; it was a deliberate decision rooted in precision and branding genius. Producer Albert R. "Cubby" Broccoli and director Guy Hamilton sought a car that could capture Bond's essence—British craftsmanship, speed, refinement, and a hint of danger. They approached several manufacturers, but only Aston Martin had the mystique that perfectly complemented Bond's persona.

The DB5, first launched in 1963, was already an evolution of the acclaimed DB4. It featured a 4.0-liter straight-six engine capable of propelling the car to 145 mph—a thrilling figure for its era. It wasn't the fastest car on the planet, but it was certainly one of the most beautiful. Every curve, every hand-sculpted aluminum panel spoke of craftsmanship and quiet aggression.

What truly immortalized the DB5, however, wasn't its speed but its *secrets*. The gadget-laden version crafted by the fictional Q Branch gave audiences a new dimension of fantasy. Revolving license plates, twin machine guns hidden behind the headlights, retractable bulletproof screens, tire slashers, and, most famously, the ejector seat—these weren't just gimmicks;

they were extensions of Bond himself. Each gadget represented wit, preparedness, and cool under pressure.

The car became a metaphor. Just as Bond's charm masked his lethal efficiency, the DB5's refined silhouette concealed a formidable arsenal. When the DB5 raced across the Alpine passes of Switzerland in *Goldfinger*, pursued by henchmen and gunfire, cinema witnessed a new kind of heroism—one expressed through design and engineering as much as through dialogue or action.

The DB5's appearance in *Goldfinger* sparked a worldwide phenomenon. Aston Martin showrooms overflowed with visitors. Children collected die-cast DB5 toys; adults dreamt of owning one. The DB5 became not only a movie star but a cultural artifact—an object of aspiration and status. It was as if every man who watched Bond drive the DB5 secretly believed he, too, could become 007, if only he had the keys.

In less than two hours of film, Aston Martin transformed from a respected marque into a global legend. The DB5 was no longer just an automobile—it was the embodiment of desire, precision, and British identity on four wheels.

After *Goldfinger* came *Thunderball* (1965), and with it, the DB5's second appearance. If the first film had introduced the car, the second solidified its myth. By now, the DB5 had transcended cinema. It had become a permanent fixture in the psyche of automotive enthusiasts, film lovers, and the public at large.

In *Thunderball*, the DB5's role was expanded, showcasing more daring stunts and creative gadgetry. The Bond franchise had realized that audiences didn't just come for the spy—they came for his car. This realization would anchor a partnership that spanned six decades.

The car returned briefly in later films such as *GoldenEye* (1995), where Pierce Brosnan's Bond revived the legend for a new generation, and again in *Skyfall* (2012), where Daniel Craig's modern, more vulnerable Bond found solace in the same iconic machine. When the DB5 reappeared in *Skyfall*, rising from its secret Scottish hideout, audiences erupted in nostalgic applause. The car's reintroduction was not mere fan service; it symbolized continuity—a reminder that some legends never die, they only evolve.

The DB5's longevity within the Bond series mirrored Aston Martin's own resilience as a company. Through financial turmoil, ownership changes, and shifting markets, the brand's identity remained unshaken. Just as Bond adapted to a new world with each generation, Aston Martin continually reinvented itself without ever losing sight of its core philosophy: grace under fire, performance under pressure.

Every reappearance of the DB5 in a Bond film served as a reminder of craftsmanship, heritage, and national pride. For many fans, the sight of the DB5's gleaming silver bodywork and the growl of its straight-six engine was not just nostalgia—it was reassurance that certain forms of excellence never fade.

Aston Martin's collaboration with the Bond franchise changed the company's trajectory forever. Before *Goldfinger*, Aston Martin was admired but relatively niche. After *Goldfinger*, it was a household name. Sales soared, media coverage expanded, and the DB5 became shorthand for elegance and power.

But the Bond Effect went beyond simple marketing. It created an emotional bond—pun intended—between Aston Martin and the public imagination. The brand was no longer just a car manufacturer; it was an emblem of adventure, masculinity, and sophistication.

From a business perspective, this was invaluable. Aston Martin's association with Bond elevated its global visibility in ways that advertising could never achieve. The car became a cultural ambassador for Britain—representing not just luxury, but resilience, innovation, and charm. In many ways, the DB5's cinematic success saved Aston Martin, helping the company weather the financial storms that would follow in subsequent decades.

This phenomenon wasn't limited to film audiences. The DB5's appearance spurred an entire merchandising empire—model cars, posters, collector's editions, and special reissues. Each generation rediscovered the magic, ensuring that the DB5 remained in perpetual motion in the public's heart.

The "Bond Effect" transformed Aston Martin into more than a car brand—it became a *mythology*. Whenever Bond drove an Aston Martin, it wasn't just a scene—it was a declaration. The message was clear: elegance and danger could coexist, and true craftsmanship was timeless.

Behind every cinematic legend lies an army of engineers, craftsmen, and dreamers who turn fantasy into mechanical reality. The DB5's transformation from luxury GT to super-spy machine required extraordinary ingenuity.

For *Goldfinger*, Aston Martin supplied two pre-production DB5s to Eon Productions. These cars were modified at Pinewood Studios and at Aston's Newport Pagnell facility, where engineers collaborated with the film's special effects team to integrate the car's now-iconic gadgets.

The rotating license plates were a marvel of 1960s mechanical improvisation—an elegant clockwork mechanism powered by a small electric motor. The hidden machine guns were actually Browning replicas, carefully wired to ignite flashes of flame through the barrels using squib charges. The ejector seat mechanism, though never fully functional for safety reasons, was an inspired piece of set engineering, using compressed air to simulate the blast.

Even the bulletproof rear shield—a combination of steel and aluminum—was meticulously fabricated to maintain the car's sleek profile. These details elevated the DB5 from prop to protagonist.

Aston Martin engineers watched in fascination as their car transcended its mechanical purpose to become cinematic history. "We didn't build it to be a movie star," one engineer reportedly said, "but it seems that's what it was born for."

Indeed, the craftsmanship that defined Aston Martin's road cars translated perfectly to the big screen. Precision, performance, and beauty—three pillars of the Aston Martin ethos—were mirrored in every frame of Bond's adventures.

By the late 1960s, Aston Martin had become a symbol of aspiration not just in Britain but across the world. The DB5's appearances in *Goldfinger* and *Thunderball* coincided with the rise of the Jet Age—a time when style, travel, and sophistication were the ultimate status symbols.

In America, sales inquiries poured in, even though Aston Martin's production numbers remained limited. The scarcity only increased desirability. Owning an Aston Martin was no longer about simply owning a car—it was about owning a *statement*.

The global fascination extended to every corner of pop culture. Magazines featured full-color spreads of Connery beside the DB5. Toy manufacturers produced miniature versions by the millions. Even children who had never seen an Aston Martin in real life could recognize it instantly by its silhouette.

For Aston Martin, this was the ultimate validation. Decades of meticulous engineering, of racing heritage, of craftsmanship—it all converged in one cinematic moment that resonated across generations. The Bond partnership gave Aston Martin a kind of immortality that even the finest race victories could not achieve.

The passage of time has only deepened the DB5's mystique. When Daniel Craig's Bond rediscovered the DB5 in *Skyfall*, parked beneath a dusty tarp in a remote garage, audiences gasped collectively. It wasn't merely nostalgia—it was reverence.

That single reveal spoke to something profound: in an age dominated by digital effects and disposable trends, authenticity still mattered. The DB5 represented permanence, craftsmanship, and heritage—values that modern audiences longed for.

Aston Martin capitalized on this timeless appeal by producing limited continuation models of the DB5, complete with authentic gadget replicas. These cars, crafted by hand at Newport Pagnell, sold for millions. Yet for collectors, the value wasn't measured in horsepower or speed, but in history. To own a DB5 was to possess a piece of cinema, of culture, of national pride.

The synergy between Bond and Aston Martin was unique because it was mutually beneficial. Bond gave Aston Martin cinematic immortality; Aston Martin gave Bond tangible sophistication. It was a perfect alignment of identity and image.

Other film franchises have tried to replicate this formula—product tie-ins, flashy cars, and tech-heavy vehicles—but none have achieved the same resonance. Why? Because the Bond-Aston Martin partnership wasn't contrived. It was authentic. Both stood for excellence born of struggle, refinement tempered by danger, and elegance under pressure.

This authenticity forged an emotional bond that transcended marketing. When audiences see Bond behind the wheel of an Aston Martin, they don't see product placement—they see destiny.

The DB5's silver gleam has reflected generations of dreams, from Connery's golden age to Craig's gritty realism. Its roar is more than mechanical—it's cultural, symbolic, eternal.

In a century defined by fleeting trends, Aston Martin's relationship with James Bond stands as a testament to endurance. It proves that when craftsmanship, artistry, and storytelling unite, they can create not just a brand or a film—but a legend.

The DB5 remains the ultimate co-star—a perfect union of man and machine. And as long as 007 continues to charm, chase, and conquer, an Aston Martin will always be waiting in the shadows, engine purring, ready for the next adventure.

As the new millennium dawned, both the Bond franchise and Aston Martin stood at crossroads. Cinema had changed; audiences demanded realism, grit, and emotional depth. Cars, too, had evolved—from mechanical masterpieces to technological ecosystems. Yet amid all this change, one truth remained unshaken: Aston Martin *was* James Bond's soul on wheels.

When *Casino Royale* (2006) premiered, the world witnessed not a reboot, but a *rebirth*. Daniel Craig's portrayal of Bond stripped the character to his essence—vulnerable, brutal, and human. And beside him stood the perfect embodiment of this evolution: the **Aston Martin DBS**.

Gone was the nostalgia of the 1960s. The new DBS was a weaponized sculpture, a blend of carbon fiber and aggression. Its 6.0-liter V12 engine unleashed 510 horsepower—raw, unfiltered power delivered with surgical precision. But beneath its ferocity lay Aston Martin's timeless DNA: balance, poise, and undeniable beauty.

The DBS didn't just appear in *Casino Royale*; it made a statement. The legendary crash scene—where Bond flips the car seven times in pursuit of Vesper Lynd—became a cinematic milestone, earning a Guinness World Record for the most barrel rolls by a vehicle in a film. It was violent, spectacular, and unforgettable. The moment symbolized Bond's vulnerability, but also Aston Martin's unwavering dominance as his chariot through chaos.

This was no longer nostalgia; it was rebirth through fire. The partnership between Bond and Aston Martin had evolved, proving that legends could adapt without losing their essence.

In *Quantum of Solace* (2008), the DBS returned—sleeker, darker, and even more menacing. Its presence in the film's opening chase sequence across Lake Garda, Italy, was nothing short of cinematic perfection. The scene encapsulated the essence of modern Bond: relentless pursuit, controlled aggression, and elegance in motion.

Unlike the gadget-laden DB5 of the 1960s, this DBS relied on pure performance. It didn't need hidden weapons or ejector seats—it *was* the weapon. The deep growl of the V12, the crisp shifts of the paddles, and the brutal choreography of the chase gave audiences a visceral reminder that Aston Martin's engineering was as breathtaking as its design.

The car's design itself spoke volumes. Aerodynamic curves replaced vintage flourishes, while its carbon-ceramic brakes and lightweight panels hinted at a company that had mastered both art and science. The DBS was not built for show—it was built for precision.

This evolution mirrored the changing tone of the Bond series. Gone were the quips and gadgets of the past; in their place stood realism and psychological depth. Aston Martin's role evolved accordingly—from fantasy to fidelity. It no longer represented escapism—it represented authenticity, control, and power under duress.

The Bond-Aston partnership had entered a new age: a digital era where luxury meant performance born of technology, not just tradition.

If *Casino Royale* and *Quantum of Solace* reinvented the myth, *Skyfall* (2012) reaffirmed it. The film marked the 50th anniversary of the Bond franchise, and with it came a resurrection—an emotional reunion between man and machine.

In one of the most unforgettable scenes in Bond history, the **DB5** returned. Hidden away in a dimly lit Scottish garage, the car emerged from beneath a dusty tarp like a memory made metal. The moment was cinematic poetry. For longtime fans, it was a reunion with an old friend; for new audiences, it was a masterclass in how legacy meets relevance.

The DB5's reappearance wasn't just fan service—it was *symbolism*. As Bond returned to his childhood home to confront his past, the DB5 represented his history, his identity, and his unbreakable link to heritage.

When the headlights flared and the familiar engine note echoed across the moors, audiences around the world cheered. Here was proof that nostalgia, when treated with respect, could be as powerful as innovation.

Director Sam Mendes understood that the car was not merely a prop—it was a time machine. Every gleam of chrome, every curve of metal, carried fifty years of cinematic memory. And when the DB5 met its fiery demise at Skyfall Lodge, it wasn't just a car exploding—it was an era bidding farewell.

Yet, as history has shown, the DB5 never truly dies. It simply waits to be reborn.

In *Spectre* (2015), Aston Martin and Eon Productions took their partnership to new heights—literally creating a car that didn't exist anywhere else in the world: the **DB10**.

This was not a production model. Only ten units were ever built exclusively for the film, hand-crafted at Aston Martin's Gaydon headquarters. The DB10 represented a bold experiment—a glimpse into the future of Aston Martin design language, foreshadowing the forthcoming DB11 and Vantage models.

Its design was razor-sharp, blending aerodynamic aggression with sculptural beauty. The proportions were unmistakably Aston—long hood, short tail, muscular stance—but the details spoke of evolution. Narrow LED headlights, deep body sculpting, and futuristic surfaces announced a new era.

In *Spectre*, the DB10 was more than a vehicle; it was Bond's second skin. Sleek, silent, and deadly, it matched his precision and poise. The car's Rome chase sequence against the Jaguar C-X75 remains one of the most visually striking pursuits in modern cinema—a ballet of speed through history-laden streets.

The DB10's real-world significance cannot be overstated. It marked a return of Aston Martin's design leadership. Under Chief Creative Officer Marek Reichman, the company redefined its aesthetic philosophy—emphasizing proportion, purity, and power. The DB10 wasn't just a movie car; it was a mission statement: *Aston Martin would never follow—it would lead.*

When one of the ten DB10s was auctioned for charity in 2016, it fetched an astounding £2.4 million. A testament, once again, to the Bond Effect's power to transform engineering into mythology.

By the time *No Time To Die* (2021) arrived, both Aston Martin and the Bond franchise had matured into global institutions. The film, Daniel Craig's swan song as 007, brought together six decades of legacy in one cinematic crescendo.

For the first time, multiple Aston Martins shared the screen: the **DB5**, **V8 Vantage**, **DBS Superleggera**, and **Valhalla** prototype—all symbols of different eras converging in a final symphony of speed.

The DB5, restored and battle-ready, opened the film with an explosive sequence in Matera, Italy—machine guns blazing from its headlights in a breathtaking homage to *Goldfinger*. This moment encapsulated what Aston Martin had become: timeless yet fierce, heritage fused with horsepower.

The V8 Vantage's reappearance, recalling Timothy Dalton's Bond in *The Living Daylights*, was a loving nod to fans. Meanwhile, the DBS Superleggera, with its 715-horsepower twin-turbo V12, showcased Aston's modern might—a grand tourer reimagined for the 21st century. And then there was the Valhalla, a hypercar prototype symbolizing the company's electric-hybrid future.

Each car carried its own story, its own era. Together, they formed a moving tribute to a relationship that had shaped culture itself.

As the film's closing scenes played out and Bond met his fate, the sight of the DB5 driving into the sunset once again stirred audiences around the world. It wasn't just a cinematic farewell—it was the sound of six decades of shared history whispering goodbye.

Few partnerships in cinema or industry have achieved what Aston Martin and Bond have accomplished together. Over sixty years, their collaboration has outlasted actors, trends, technologies, and even corporate eras.

The Bond Effect transcended film. It seeped into fashion, design, and psychology. Men dressed like Bond; women admired the confidence he exuded; car enthusiasts aspired to the sophistication embodied by his Aston. The cars became shorthand for *taste refined by danger*.

In marketing terms, it's an unmatched phenomenon. The Bond franchise provided Aston Martin with global exposure that billions of dollars in advertising could never replicate. Yet the secret wasn't visibility—it was alignment. Both brands shared the same DNA: craftsmanship, British identity, and fearless innovation.

The DB5 and its successors became cultural ambassadors. They appeared in video games, exhibitions, advertisements, and even museums. The line between reality and fiction blurred; owning an Aston Martin was no longer about transportation—it was about participating in a legend.

Even the rise of digital culture couldn't diminish the mystique. In an era when virtual cars dominate video games and electric vehicles redefine performance, the image of a Bond-era Aston Martin remains the ultimate symbol of analog passion—a reminder of the romance of the road.

Behind every cinematic roar lies a calculated business strategy. Aston Martin, often navigating financial turbulence, has long leveraged its Bond connection not as a crutch but as a compass.

From the 1960s to today, the brand has used Bond films as springboards for innovation and visibility. Each cinematic appearance coincided with key product launches, drawing global attention. For instance, the DB10 in *Spectre* previewed the DB11's design DNA; the DBS in *Casino Royale* heralded a new generation of performance GTs.

Even more importantly, Aston Martin learned to treat its cinematic presence as storytelling rather than sponsorship. The cars were characters, not commercials. This narrative approach elevated the brand above mere marketing, embedding it in emotional memory.

Collectors and investors responded. Each limited-edition "Bond Edition" model—from the DBS to the Vantage—sold out within hours. Prices on the secondary market soared, confirming that Aston Martin had achieved what few brands could: transforming fiction into financial value.

The Bond Effect became an economic ecosystem of its own—driving merchandise, tourism, events, and cross-industry collaborations. From Omega watches to Heineken campaigns, every Bond partnership reinforced the aura of refinement and daring that Aston Martin embodied.

In essence, the Bond films didn't just sell cars—they sold dreams.

Aston Martin's genius lies not only in building fast cars but in crafting emotional experiences. Every Bond-era model—from DB5 to DBS Superleggera—evokes a visceral connection that transcends performance metrics.

The sound of an Aston Martin engine isn't just acoustics—it's storytelling. The deep growl of a V12 or the refined hum of a V8 carries decades of passion, craftsmanship, and legacy.

Inside, the cabins are sculpted like the interiors of a Savile Row suit—meticulously tailored, tactile, and timeless. The use of hand-stitched leather, bespoke metals, and natural materials creates a sensory symphony that few manufacturers can replicate.

In this sense, Aston Martin has always built *Bond cars*—even outside the films. Each model carries the spirit of espionage, the allure of risk, the promise of escape. It's not just about driving; it's about *being*.

This emotional engineering is why the Bond connection remains potent even in an era dominated by sustainability and electrification. The cars evoke a primal human desire—to command, to explore, to defy limits. And as long as humans dream, that desire will never fade.

As the world moves toward electrification, the Bond-Aston Martin partnership faces a thrilling new challenge: how to preserve soul in a silent world.

Aston Martin's upcoming Valhalla hybrid and all-electric models like the Rapide E hint at an evolution of performance philosophy. The brand is committed to crafting electric vehicles that retain the visceral emotion of combustion while embracing sustainability.

Will Bond drive an electric Aston in the next era? Almost certainly—and when he does, it will redefine cool all over again. Because what matters isn't the roar of the engine, but the heartbeat of the legend.

In the electric age, design and storytelling will matter more than ever. Aston Martin's future cars will carry not only advanced battery systems and AI-driven dynamics but also the DNA of the DB5—the same craftsmanship, elegance, and power that have defined six decades of 007's adventures.

The stage is set for a new chapter: one where technology and tradition, sustainability and seduction, coexist in perfect tension.

The story of James Bond and Aston Martin is not just about cars or cinema—it's about *identity*. It's about how two British icons became reflections of one another, evolving with time but never surrendering their essence.

From the silver DB5 of *Goldfinger* to the carbon-clad Valhalla of tomorrow, their partnership has spanned sixty years of artistry, innovation, and imagination.

The DB5 may have been born in 1963, but it still commands hearts in 2025. It is proof that when design meets destiny, when craftsmanship meets charisma, a machine can transcend its mechanics and enter the realm of mythology.

Bond will change. The world will change. But when the next 007 steps into the light, the faint purr of an Aston Martin engine will echo once more, reminding us that legends are not written—they are *driven*.

5: THE CLASSIC V8 VANTAGE

The 1970s were turbulent yet transformative years for the automotive world. The oil crisis of 1973, the tightening of emission standards, and the shift toward more practical vehicles forced even the proudest marques to reconsider their philosophies. Yet amid this uncertainty, Aston Martin — the small British manufacturer whose cars embodied prestige and performance — decided to defy convention.

Aston Martin's engineers and designers understood that the world did not need another car. What it needed was *a statement of intent*, a machine that would remind enthusiasts what true craftsmanship and passion could achieve when the world was turning to austerity. That machine was the **Aston Martin V8 Vantage**, launched in 1977 — a car that didn't just signal the dawn of a new generation, but redefined what an English grand tourer could be.

The V8 Vantage was not built to follow trends; it was designed to **set them**. It marked the shift from the smooth, almost aristocratic elegance of the DB series to a bolder, more muscular, and more modern identity. This was Aston Martin's answer to the Italian exotics — Ferraris and Lamborghinis that shouted for attention — but with the unmistakable restraint and refinement that only the British could master.

When the first V8 Vantage rolled out of the factory at Newport Pagnell, it was more than a car — it was a declaration. It told the world that even in an age of economic caution and corporate takeovers, Aston Martin still knew how to craft a machine that stirred the soul and commanded the road.

The DB series had been the car of gentlemen — the V8 Vantage was the car of **connoisseurs**. It was not for everyone. It was powerful, rare, and unapologetically ambitious — a car that demanded respect and rewarded those who truly understood driving.

The lineage of the V8 Vantage can be traced back to the **DBS**, which first appeared in 1967. The DBS was meant to carry Aston Martin into a new era, with its clean lines and larger dimensions, but its six-cylinder engine left some enthusiasts yearning for more. The company knew that the DBS's bold chassis and aerodynamic body deserved something far more potent beneath its bonnet.

By the early 1970s, the 5.3-liter V8 engine designed by **Tadeusz "Tadek" Marek** — an engineering genius who had earlier designed the DB4 GT's engine — became the beating heart of Aston Martin's performance ambitions. The DBS V8, introduced in 1969, was a technological marvel, but it was only the beginning.

As the decade progressed, Aston Martin's engineers began working on something far more radical — a car that could combine the grace of a grand tourer with the ferocity of a race car. That vision became the **V8 Vantage** — the first true "supercar" from Britain's most exclusive manufacturer.

The design brief was simple yet audacious: create the most powerful production Aston Martin ever made. It had to be faster, sharper, and more aggressive, yet maintain the sophistication that defined the brand. The team achieved this balance through meticulous engineering, aerodynamic refinement, and an obsessive attention to detail that bordered on artistry.

One look at the Aston Martin V8 Vantage and it was clear — this was not a car that whispered. It **announced itself**.

The shape was purposeful — a striking blend of athletic muscle and tailored refinement. Every curve, every crease in the metalwork seemed to convey confidence and poise. Gone were the gentle contours of the DB6; in their place stood wide, flared arches that embraced the car's broad tires and low stance.

The nose was sharp, almost predatory, with a wide mesh grille and twin headlamps that glared with intent. The long, sculpted bonnet stretched toward the horizon, suggesting immense power lurking beneath. Yet there was no excess, no flamboyance — just a quiet menace wrapped in elegance.

The side profile told another story: a perfect equilibrium between proportion and purpose. The long hood flowed seamlessly into a sleek cabin, tapering into a short, decisive tail. The fastback design hinted at aerodynamics, but also maintained Aston Martin's classic grand touring silhouette — a nod to tradition wrapped in a modern suit.

At the rear, the car exuded power through subtle details — quad exhaust pipes, a sculpted boot lid, and carefully integrated lights. The overall design felt balanced and deliberate, every inch designed to project authority.

Inside, the V8 Vantage retained the old-world charm of British luxury. Hand-stitched Connolly leather adorned the seats, the dashboard was trimmed with polished walnut veneer, and every switch and dial felt tactile, mechanical, and precise. It was a driver's environment — intimate, crafted, and exuding a sense of permanence.

To drive a V8 Vantage was to sit in a **handmade masterpiece**, not a mass-produced machine. Even the smell — that unique blend of leather, wood, and oil — spoke of authenticity.

Under its long bonnet, the V8 Vantage hid a piece of engineering that redefined the company's capabilities — the **5.3-liter V8 engine**, tuned to produce an astonishing 380 horsepower in later versions. This made it one of the most powerful production cars in the world at the time.

For perspective, this was an era when even the Ferrari 308 produced around 255 horsepower. The Aston Martin V8 Vantage was not just keeping up — it was *leading*.

The car's performance figures were staggering for the late 1970s: 0 to 60 mph in 5.2 seconds, and a top speed approaching 170 mph. These were numbers that placed it in the same league as the world's most exotic supercars, yet the V8 Vantage achieved them while maintaining the civility of a gentleman's grand tourer.

Aston Martin's engineers achieved this balance through meticulous tuning. The V8 engine featured **Weber carburetors**, high-lift camshafts, larger valves, and improved compression ratios. The suspension was fully independent, and the car rode on specially developed Avon tires to handle the immense torque.

The result was astonishing — a car that felt planted, responsive, and alive. Its steering was heavy at low speeds but perfectly weighted at high velocity, allowing the driver to carve through corners with precision. The ride was firm but never harsh — a reminder that the V8 Vantage was not a track toy but a refined performer built for the open road.

The braking system, too, was engineered with purpose. Ventilated disc brakes on all four wheels ensured reliable stopping power, while the weight distribution gave the car an inherently balanced feel. On winding roads, the V8 Vantage danced — not delicately like a Ferrari, but with a confidence born of brute strength and control.

Aston Martin had managed to create something that few manufacturers ever achieve — a car that combined *emotion and engineering* in perfect harmony.

The brilliance of the V8 Vantage lay not just in its numbers but in its philosophy. It was **luxury and performance fused together** in a way few cars before it had achieved.

Inside, the cabin was a sanctuary of refinement. Deep, plush leather seats hugged the driver, offering both comfort and control. The wood veneer dashboard reflected the light in subtle glows, and the analog dials — large, mechanical, and clear — gave the sense of piloting a machine with purpose.

Unlike the stripped-down sports cars of its era, the V8 Vantage never compromised on comfort. Air conditioning, high-quality sound systems, and even power windows were part of its offering — rare luxuries in a high-performance car at the time.

But what made the V8 Vantage truly special was how it *felt* on the road. The engine didn't just produce power — it produced character. The deep, thunderous roar of the V8 was intoxicating, a symphony of mechanical precision that rose with each press of the throttle.

Driving a V8 Vantage was not about chasing lap times. It was about **presence**. It was about the way the car made its driver feel — powerful, composed, and unmistakably distinguished. The world outside blurred, and for that brief stretch of road, you weren't just driving — you were commanding.

In 1977, when the V8 Vantage debuted, it was hailed as "Britain's first true supercar." This was not just marketing hyperbole — it was a statement of fact.

While Italy had its Ferraris and Lamborghinis, and Germany had its Porsches, Aston Martin had created something entirely different. The V8 Vantage was not about flamboyance or razor-sharp minimalism — it was about **grand authority**. It was a car that didn't need to scream to be noticed; it simply exuded presence.

Car magazines of the era were astonished by its performance. *Autocar* famously tested the car and declared it "a British thoroughbred that can outrun the world's best." Critics praised its balance, its refinement, and its sheer sense of occasion.

Even more impressive was that every V8 Vantage was built almost entirely by hand at the Newport Pagnell factory. Each car took hundreds of hours to assemble, and no two were ever truly identical. It was this artisanal process — rare even then — that gave each V8 Vantage its unique soul.

The result was a car that could sprint like a race machine yet cruise across Europe in comfort. It wasn't merely transportation; it was **motoring art** — the embodiment of Aston Martin's motto: *Power, Beauty, Soul.*

To truly understand the V8 Vantage, one must understand the time it was born into. The 1970s were not kind to performance cars. Fuel shortages, environmental regulations, and economic recessions had forced many manufacturers to downsize or abandon their high-performance ambitions altogether.

Aston Martin, however, refused to compromise. The company's leadership — despite its financial struggles — believed that performance and elegance could coexist. The V8 Vantage became a symbol of **resistance**, a defiant roar in an age of whispers.

Its bold design, thunderous engine, and impeccable craftsmanship made it a statement piece — not just for the company, but for every enthusiast who believed driving was an art form.

For Aston Martin, the V8 Vantage was survival through excellence. It kept the brand alive at a time when others faltered. More than that, it redefined what a grand touring car could be.

By the time 1987 arrived — a full decade after its debut — the Aston Martin V8 Vantage had already achieved legendary status. But fate had another chapter planned.

That year, **James Bond** would once again slide behind the wheel of an Aston Martin — this time, the very same V8 Vantage. The car's appearance in *The Living Daylights* transformed it from an automotive masterpiece into a global icon.

It was more than product placement; it was destiny fulfilled. The Bond franchise had always been entwined with

Aston Martin's identity, and now the V8 Vantage carried that torch into a new era.

It is impossible to speak of the Aston Martin V8 Vantage without invoking the shadow of the world's most famous secret agent — **James Bond**. For decades, Aston Martin and Bond had shared a relationship defined by elegance, sophistication, and sheer performance. But when the V8 Vantage appeared in *The Living Daylights* (1987), the partnership reached a new level of cinematic and cultural immortality.

When Timothy Dalton took over the role of 007, the producers sought a Bond who was more grounded — serious, dangerous, and human. It was only fitting that his car should reflect the same transformation: sleek, understated, and devastatingly effective. Enter the Aston Martin V8 Vantage.

In the film, Bond's V8 wasn't just a vehicle — it was a weapon. Outfitted with rocket propulsion, retractable skis, missile launchers, lasers, and even a self-destruct system, it was a perfect blend of style and lethality. Though the gadgets were fictional, they were rooted in a car that already embodied power and precision in real life.

For millions of viewers, this was their first introduction to the V8 Vantage — and what an introduction it was. The car's dark metallic paint gleamed under the Austrian snow as Bond outmaneuvered enemy agents, skidding across frozen lakes and blasting through barriers with ruthless grace. The V8 Vantage wasn't merely a supporting character; it *stole the scene*.

The film cemented the V8 Vantage's status as the "**James Bond car for a new generation**." While Sean Connery's DB5 had symbolized suave sophistication in the 1960s, Dalton's V8

Vantage embodied cold precision and raw muscle — perfectly mirroring the evolution of Bond himself.

From that moment, the V8 Vantage transcended its identity as a car. It became a symbol — of British strength, elegance, and indomitable cool. Even decades later, fans still regard it as one of the most memorable Bond vehicles ever filmed.

The car that dazzled audiences on screen was not simply a prop; it was built on a foundation of uncompromising engineering. Underneath its cinematic charm lay the mechanical DNA that had made the V8 Vantage one of the fastest, most powerful grand tourers of its era.

The heart of the car remained its **5.3-liter V8 engine**, meticulously tuned to produce between 380 and 400 horsepower — numbers that rivalled Ferraris and Lamborghinis of the day. The torque delivery was immense, peaking at around 400 lb-ft, providing an effortless surge of power in any gear.

This engine wasn't just about brute force. It was about **refinement**. Engineers spent countless hours perfecting the sound and feel of the car. The throttle response was instant, the exhaust note a deep, resonant growl that could turn into a full-throated roar at the flick of a right foot.

Aston Martin's commitment to hand-built craftsmanship ensured that each V8 Vantage was unique. Every engine was individually assembled and tested by a single engineer, whose name was often discreetly stamped onto a plaque — a mark of pride and accountability. This practice, now common in luxury performance brands, was revolutionary in its time.

The V8 Vantage's **chassis and suspension** setup also played a crucial role in its success. While its Italian rivals favored lightness and razor-sharp handling, Aston Martin pursued stability and confidence. Double-wishbone suspension, coil springs, and anti-roll bars ensured a poised, predictable ride even at high speeds.

Behind the wheel, the car felt planted — solid, commanding, and immovable. It inspired trust, giving the driver an uncanny sense of control, whether gliding along the Autobahn or tackling British B-roads in the rain.

The combination of engineering excellence and design mastery meant that the V8 Vantage could deliver thrilling performance without compromising the ride quality that Aston Martin owners cherished. It was not a car that demanded constant correction; it was a car that invited you to *become one with the drive*.

One of the most defining characteristics of the V8 Vantage was that it was **entirely handmade** — a rarity even then, and almost unheard of today.

At the Newport Pagnell factory in Buckinghamshire, the air was filled with the scent of leather, metal, and lacquer. Here, craftsmen shaped each body panel by hand, using wooden bucks and precision tools. Every car was built by a small team of artisans who treated their work not as manufacturing, but as **artistry**.

There were no assembly lines in the modern sense. Instead, each car moved slowly through the factory, evolving from bare chassis to finished masterpiece over hundreds of man-hours. The process was painstaking but deeply personal. Engineers knew every bolt, stitch, and curve by touch.

The interior craftsmanship was equally extraordinary. The leather was sourced from the finest hides, hand-cut and stitched to fit perfectly around the seats and dashboard. Wood veneers were polished and matched for grain consistency. The attention to detail bordered on obsession — even the stitching direction on the steering wheel was chosen for comfort and aesthetic balance.

Customers who ordered a V8 Vantage were not merely buying a car — they were commissioning a piece of **rolling British heritage**. Each vehicle was built to order, tailored to the owner's specifications, whether that meant bespoke color palettes, special interior trims, or performance enhancements.

This devotion to craftsmanship gave every V8 Vantage a soul. It was not just a machine; it was the culmination of hundreds of hours of human labor, care, and pride.

As the 1980s gave way to the 1990s, the automotive landscape was shifting again. The rise of computer-controlled fuel injection, turbocharging, and mass production was transforming how performance cars were built. Yet the V8 Vantage remained defiantly analog — a last bastion of **mechanical purity**.

Driving the V8 Vantage was an experience that demanded engagement. The steering was heavy, the clutch firm, and the gearbox mechanical. It was a car that asked for respect — and rewarded it tenfold.

This analog quality made it even more appealing to purists. In an age of increasing automation, the V8 Vantage reminded drivers of what motoring truly meant: the connection between man and machine, the sound of pistons firing, and the feeling of raw power coursing through the chassis.

Aston Martin's decision to continue building cars by hand also meant that production numbers remained low. Between 1977 and 1989, only a few hundred V8 Vantages were ever made, making each one exceptionally rare.

Collectors soon realized the car's significance. By the early 2000s, well-preserved examples were fetching prices that far exceeded their original cost. Today, a pristine V8 Vantage commands six-figure sums, and its value continues to rise — a testament to its timeless appeal.

In 1986, Aston Martin unveiled a new variant that would become just as iconic — the **V8 Vantage Volante**, a convertible version that blended open-air freedom with breathtaking power.

The Volante captured the essence of grand touring in its purest form. With the same 5.3-liter V8 engine and muscular styling, it combined the Vantage's performance with an added layer of drama and glamour.

The car's proportions — long hood, short tail, and wide stance — translated beautifully into a convertible format. Whether cruising through the English countryside or along the French Riviera, the Volante exuded effortless sophistication.

The model achieved international fame when it too appeared in *The Living Daylights*. Bond's Volante, with its roof removed and gadgets engaged, became the ultimate fantasy of freedom and danger. It wasn't just a car anymore; it was a **cultural icon** — the embodiment of British elegance and adventure.

Today, the V8 Vantage Volante remains one of the most collectible Aston Martins ever made, prized for its rarity, craftsmanship, and cinematic heritage.

Every great car leaves a mark — and the V8 Vantage's influence on Aston Martin's future was profound.

When the company reimagined its lineup in the early 2000s, the **modern V8 Vantage (2005)** drew heavily from the DNA of its predecessor. The proportions, the stance, and even the philosophy — performance wrapped in sophistication — all traced their lineage back to the 1977 classic.

Aston Martin's designers openly acknowledged this heritage. The muscular fenders, the assertive grille, and the blend of athleticism and luxury were direct descendants of the original. The very spirit of the brand — that delicate balance of aggression and grace — was defined by the V8 Vantage and carried forward into every model thereafter.

Even in today's era of electrification and digital engineering, the echoes of the 1970s Vantage can still be felt. It serves as a reminder that **true luxury is timeless**, and that emotion and craftsmanship cannot be replicated by technology alone.

The V8 Vantage was more than a car; it was a philosophy. It proved that elegance and power could coexist, that a grand tourer could thrill as much as it soothed, and that the human touch was irreplaceable in the pursuit of excellence.

Beyond engineering and design, the V8 Vantage became a cultural symbol — of success, taste, and individuality. It wasn't ostentatious like a Ferrari, nor clinical like a Porsche. It was a car for those who understood restraint — who preferred quiet confidence over loud statements.

In Britain, it represented **national pride**. While many luxury marques fell to foreign ownership or faded into obscurity, Aston Martin and its V8 Vantage stood firm as symbols of homegrown excellence. The car embodied everything quintessentially British: craftsmanship, charm, and defiance.

In film and media, its presence endured. Even in 2021's *No Time to Die*, the V8 Vantage made a triumphant return to the Bond franchise, now driven by Daniel Craig's 007 — a full-circle moment that bridged five decades of cinematic and automotive history. As the camera lingered on the car's sleek lines and rumbling engine note, it was clear: legends never die; they evolve.

The crowd's reaction to its reappearance was immediate and visceral. For enthusiasts, it was not just nostalgia — it was recognition. The V8 Vantage wasn't a relic; it was a living testament to enduring design and engineering excellence.

To drive an Aston Martin V8 Vantage today is to experience a piece of history that feels alive. The car demands engagement, rewards skill, and offers sensations that modern vehicles — no matter how advanced — often fail to replicate.

Each drive is an event. The sound of the V8 firing to life, the vibration through the leather-clad steering wheel, the view over the long hood — it all speaks to an era when cars were crafted, not produced.

Collectors describe the experience as **spiritual**. The car communicates with its driver in ways that transcend numbers and specifications. Its imperfections — the weight of the clutch, the mechanical thrum of the engine — are not flaws, but features that give it personality.

This is what sets the V8 Vantage apart from other classics. It doesn't just look beautiful; it feels *authentic*. It carries the spirit of the craftsmen who built it and the drivers who pushed it to its limits.

The story of the Aston Martin V8 Vantage is not merely about horsepower, speed, or fame. It is a story of **identity** — of a brand that refused to compromise, of engineers who refused to follow trends, and of a car that became an icon against all odds.

It was born in adversity, matured in glory, and lives on in legend. It bridged eras — from the classical charm of the DB series to the modern performance of the 21st century — all while maintaining an unwavering commitment to craftsmanship and style.

In an industry often driven by technology and efficiency, the V8 Vantage remains a reminder that the heart of a great car lies not in its computers, but in its *soul*.

When its engine starts, it doesn't just make noise — it tells a story. Of artisans and dreamers. Of roads and revolutions. Of one of the greatest cars ever to wear the Aston Martin wings.

The **Aston Martin V8 Vantage** will forever stand as the definitive symbol of **Power, Beauty, and Soul** — the golden trinity that defines not just a brand, but a belief that driving is, and always will be, an art.

6: ICONIC DESIGN PHILOSOPHY

There are few marques in the automotive world that command the same reverence as Aston Martin. For over a century, this British icon has balanced performance and prestige with a visual sophistication that transcends time. The brand's cars are not just machines—they are sculptures in motion, designed with a precision that borders on obsession. Every line, curve, and contour speaks of heritage, while every evolution whispers innovation.

From the earliest pre-war racers to the sleek, thunderous grand tourers of today, Aston Martin has developed a **design language unlike any other.** Its vehicles exude strength without aggression, power without arrogance, elegance without fragility. They are, as company designers often say, "beautiful machines designed for beautiful drivers." The fusion of engineering and artistry lies at the heart of this philosophy—a principle that has guided every Aston Martin since its inception in 1913.

As we explore this philosophy, we uncover how Aston Martin turned design into an identity. The company's creative journey has not merely followed the trends of the times—it has often set them, influencing generations of carmakers who sought to capture a fraction of its visual poetry.

In its earliest years, Aston Martin's designs reflected the engineering priorities of the age. Cars such as the **Aston Martin A3** and **DB1** were functional, stripped-down machines built primarily for racing and mechanical excellence. The lines were clean but practical; the bodywork was crafted for aerodynamic efficiency rather than allure. Yet, even then, an undercurrent of aesthetic purpose flowed beneath the metal.

The founders, Lionel Martin and Robert Bamford, believed that a car should not only perform well—it should *look* capable of such performance. This principle quietly seeded what would later blossom into Aston Martin's signature design language: a harmony between speed, proportion, and artistry.

By the 1930s, when the company began producing sports tourers, the cars displayed a distinct British charm. Long hoods, sweeping fenders, and neatly balanced profiles began to define their form. These early silhouettes foreshadowed the coming of a new era—one where Aston Martin would elevate car design into an art form.

The defining chapter of Aston Martin's visual identity began in the 1950s with the **DB2**, the first model to carry the initials of then-owner David Brown. The DB series did more than revive the company—it gave it a **face**. With its smooth lines, curvaceous flanks, and poised stance, the DB2 introduced a new visual standard that would carry through decades.

Then came the **DB5**, the car that changed everything. Its proportions were immaculate—long bonnet, short tail, and a waistline that seemed sculpted by wind and elegance itself. The chrome grille became its signature smile, and the round headlights, like the eyes of a refined predator, gave the car an unmistakable personality. The DB5 was not only a symbol of performance—it was a **masterclass in timeless automotive design**.

When James Bond drove the DB5 onto cinema screens in *Goldfinger*, the world collectively gasped. It wasn't just a car; it was a style statement, a cultural icon, a symbol of British sophistication. From that moment, Aston Martin's design language was forever tied to the ideals of luxury, intrigue, and power.

The following decades saw Aston Martin refine and experiment. By the time the **V8 Vantage** roared into the 1970s, the company's design team—led by masters like William Towns—embraced a more muscular and angular aesthetic. Gone were the soft, flowing curves of the DB series. In their place came squared-off headlights, flared arches, and a stance that radiated dominance.

The shift wasn't a departure from elegance—it was an evolution of it. The 1970s demanded a car that mirrored the era's boldness, and the V8 Vantage delivered precisely that. It was the automotive equivalent of a Savile Row suit tailored for a rock star: sharp, confident, and unflinchingly British.

This transformation also demonstrated Aston Martin's **design adaptability**. The company never allowed its visual identity to stagnate. Instead, it reinterpreted elegance to fit the spirit of each decade, balancing innovation with restraint. The V8 Vantage carried the same DNA as the DB5, but it spoke in a more assertive tone—a reflection of both progress and continuity.

What makes an Aston Martin instantly recognizable even at a glance? The answer lies in a few **defining design**

signatures—visual cues that have been refined and reimagined over time, yet remain unmistakably Aston Martin.

1. The Iconic Grille – Power Framed in Elegance

The Aston Martin grille is more than a design element—it is the soul of the car's face. Often described as "the company's heartbeat in metal," its distinct shape has remained the brand's most enduring symbol.

From the vertical-slat grilles of early racers to the wide, confident "smiling" grille of the DB5, its evolution tells the story of Aston Martin's growth. The grille's curvature, typically accentuated by a polished surround, gives each model a unique personality while maintaining a common lineage. In recent years, as with the **DB11** and **Vantage**, the grille has grown larger and more pronounced—an intentional choice to give modern models a bold, athletic presence.

Beyond its aesthetics, the grille serves a critical aerodynamic and thermal purpose, channeling airflow to cool the V8 or V12 engines beneath. Yet, Aston Martin has always designed it as an artful signature, ensuring it complements the car's emotional appeal.

2. The Headlights – Precision with Personality

Aston Martin's headlights tell a story of evolution—from the **circular eyes of the DB5** to the **sleek, blade-like lenses** of modern models. These lights are not just functional; they define the car's attitude.

Where earlier designs reflected innocence and charm, today's headlights embody precision and determination. The integration of advanced LED and laser technology allows for thin, sharply contoured units that enhance both aesthetics and performance. Designers often describe the headlights as "the eyes of the car"—they express character, emotion, and intent.

What makes Aston Martin's approach unique is **integration**. The headlights never feel separate from the car; they are woven into the body lines, flowing seamlessly into the fenders and curvature. This harmony gives every Aston Martin its sculptural cohesion, making it appear as if it were hewn from a single block of metal.

3. Sculpted Body Contours – Motion in Stillness

One of Aston Martin's most breathtaking qualities is how it **captures motion even when stationary**. Every curve, crease, and contour is deliberate, guiding both the eye and the wind. The car's surfaces appear alive, shifting subtly as light glides across the paintwork.

This fluidity isn't accidental—it's the product of countless hours of wind-tunnel refinement and clay modeling. The designers aim for what they call "controlled tension," where each line serves both a visual and aerodynamic purpose.

A perfect example lies in the **side strake**, the vent just behind the front wheel arch. Originally functional for cooling and aerodynamics, it became a visual hallmark—appearing in models from the DB4 to the modern Vantage. The strake is Aston Martin's equivalent of an artist's signature flourish.

Even the rear, often overlooked in other brands, receives meticulous attention. From the rounded tails of the DB5 to the bold, blade-like taillights of the DBS Superleggera, the rear design ensures that the car is as beautiful leaving as it is approaching.

4. The Proportion Principle – The Golden Ratio on Wheels

Aston Martin's designers adhere closely to one sacred law: **proportion is perfection**. The brand's hallmark long-hood, short-deck layout, combined with a cabin set far back on the chassis, evokes a sense of dynamic readiness. The car always looks poised to leap forward—an aesthetic born from racing but refined for luxury.

This principle dates back to the early DB models, where engineers sought to accommodate large engines up front without compromising visual harmony. The result was a silhouette that exuded both grace and aggression—a balance so perfect that it became an industry benchmark.

Even as modern safety and aerodynamic regulations have altered vehicle proportions, Aston Martin continues to retain this **golden ratio**. Subtle tweaks in curvature and line tension ensure that every model—no matter how futuristic—still carries that unmistakable Aston Martin stance.

5. Craftsmanship in Detail – The Human Touch

In an era dominated by automation, Aston Martin remains deeply **human** in its approach to design and construction. Every panel is hand-fitted, every stitch of leather hand-sewn. Designers collaborate directly with engineers and craftsmen to ensure the aesthetic vision translates seamlessly into the final form.

This meticulous attention to detail can be seen in the stitching patterns, the grain of the wood veneer, and the way metal trim reflects ambient light. It's no exaggeration to say that each car carries a fingerprint of the people who built it.

As design chief Marek Reichman often says, *"You feel an Aston Martin, not just see it."* That philosophy separates the brand from those who design by computer. Every Aston

Martin has a soul—a sense of craftsmanship that transforms driving into an experience of artistry.

Few brands have exerted as much aesthetic influence on the automotive industry as Aston Martin. Across continents and decades, the company's design ethos has inspired generations of designers and redefined how luxury performance cars should look. Aston Martin did not follow design movements—it created them.

In the 1950s and 1960s, when Italian marques like Ferrari and Maserati leaned heavily into flamboyance, Aston Martin offered an alternative vision—**understated power, tailored elegance, and functional beauty**. The British approach resonated deeply with a clientele that valued refinement over ostentation. This quiet confidence became an Aston Martin trademark and a point of inspiration for other automakers worldwide.

Luxury manufacturers began to study Aston Martin's proportions, stance, and use of surface tension. Even modern sports sedans and grand tourers owe aspects of their form to the DB series' balanced proportions and the V8 Vantage's muscular grace. The ripple effect of this design legacy can be seen not only in performance brands but also in mainstream vehicles striving for a sense of prestige.

But Aston Martin's influence wasn't limited to aesthetics—it redefined how design could communicate personality. Every Aston Martin was imbued with emotion. The vehicles weren't designed just to perform, but to **evoke desire, presence, and a sense of belonging**. They became expressions of individuality—machines that made statements without ever needing to shout.

Entering the 21st century, Aston Martin faced a critical question: how do you evolve a design language built on timelessness without losing its soul? The brand's answer was simple yet profound—**innovation through emotional continuity**.

Under design director **Marek Reichman**, Aston Martin reinterpreted its heritage for a new generation. The DB9, introduced in 2004, was a turning point—a masterful blend of classic cues and modern sophistication. Its bodywork was taut, sculpted, and organic, flowing like liquid metal. It wasn't just designed; it was **shaped with feeling**.

Reichman and his team coined a phrase that has since become central to the company's ethos: *"The art of beauty through proportion."* Every surface of an Aston Martin is designed to evoke emotion, to create a connection between human and machine. Unlike many manufacturers that design from the outside in, Aston Martin's process begins with emotion—what the driver should feel when seeing, touching, and driving the car.

This philosophy birthed cars like the **DB11** and **Vantage**, each showcasing a perfect balance between heritage and modernity. The DB11's floating roofline, for example, was not merely a design flourish—it visually lightened the car's silhouette while subtly improving aerodynamics. Similarly, the new Vantage's muscular haunches and predatory stance conveyed raw power without excess aggression.

Aston Martin's design approach evolved beyond simply creating cars—it became **sculptural storytelling**. Every crease told a story of tension and release; every surface reflected purpose. These weren't just vehicles; they were living forms of expression.

In today's hyper-digital world of computational design and efficiency-obsessed engineering, many cars risk losing their emotional warmth. Aston Martin has taken a defiant stance—embracing technology while preserving its design soul. The company's challenge in the 2020s and beyond is not just to make cars faster or lighter but to ensure they **continue to move people emotionally**.

One of the greatest innovations in recent years has been **Aston Martin's use of airflow as a design element**. In cars like the DB11 and DBS Superleggera, the designers introduced the concept of "AeroBlade" technology—air channeled through hidden vents that replace traditional spoilers, maintaining the car's clean lines while enhancing downforce. The result: aerodynamics that serve the art of design rather than distort it.

This union of science and beauty is emblematic of Aston Martin's next era. Form and function no longer exist in tension—they collaborate. Every vent, crease, and curvature performs dual duty: it looks exquisite and enhances performance.

Even materials play a role in this evolving aesthetic. Lightweight carbon fiber, aluminum, and advanced composites allow designers to sculpt more freely, reducing mass while maintaining the brand's signature muscularity. The result is an aesthetic that feels lighter, sharper, and more alive—what the brand calls "athletic elegance."

Looking forward, the advent of electrification brings both challenges and opportunities. Without a roaring V12 or V8 to define proportion and venting, Aston Martin's design team must rethink visual drama for the electric era. But even in this new frontier, one thing remains constant: **emotion is the core engine**. Whether powered by petrol or electrons, an Aston Martin must still quicken the pulse.

What separates Aston Martin from fleeting automotive trends is its **continuity of character**. While designs evolve, the emotional DNA remains intact. Look at any Aston Martin—from the DB2 to the modern DB12—and you will see the same spirit woven through time. The long bonnet, the low-slung cockpit, the haunches that hint at coiled energy—all whisper of the same ancestry.

This is not coincidence but deliberate craftsmanship. Designers at Aston Martin treat heritage not as nostalgia but as **living inspiration**. In the design studio, sketches of past models sit beside digital renders of future concepts. Clay sculptors reference the curvature of a 1960s DB while modeling a 2020s Vantage. This dialogue across generations ensures that the essence of Aston Martin is never lost, only reinterpreted.

Aston Martin's ability to evolve without dilution is a rare achievement in the automotive world. Brands often struggle between heritage and innovation, but Aston Martin thrives on that tension. Its design language has learned to breathe—shifting tone, refining shape, yet always speaking the same language of elegance and purpose.

Aston Martin's design influence now extends far beyond automobiles. The company has established **Aston Martin Design**, a creative studio responsible for crafting luxury experiences across industries—from architecture to aviation.

Collaborations like the **Aston Martin Residences in Miami**, the **Airbus ACH130 Aston Martin Edition helicopter**, and even the **Aston Martin AMB 001 motorcycle** showcase how the brand's aesthetic transcends metal and wheels. The same proportion principles, material craftsmanship, and emotional resonance are applied whether designing a skyscraper lobby or a pilot's cockpit.

This expansion reinforces Aston Martin's belief that **design is a universal language of emotion**. The philosophy that shaped the DB5's elegance now informs the lines of furniture, homes, and yachts. Each project carries the same design DNA: British restraint, timeless beauty, and modern dynamism. It's a bold declaration that Aston Martin is not merely an automaker—it's a curator of lifestyle.

In a world obsessed with novelty, Aston Martin's enduring appeal lies in **its commitment to restraint**. Trends may come and go—oversized grilles, exaggerated spoilers, hyper-futuristic gimmicks—but Aston Martin remains grounded in elegance.

This restraint doesn't stem from conservatism but from confidence. The company's designers understand that true beauty doesn't need to shout. Subtlety, proportion, and balance are their weapons of seduction. Where others rely on excess, Aston Martin relies on **discipline**—a rare and powerful aesthetic quality.

Timeless design also means long-term value. Unlike many sports cars that age quickly as fashions change, an Aston Martin maintains its grace through decades. The DB5 still turns heads today not out of nostalgia but because it was designed with enduring proportions. The same principle guides every modern Aston Martin, ensuring that they, too, will be admired for generations to come.

Ultimately, Aston Martin's design philosophy is not about metal, measurements, or metrics—it's about **emotion**. The brand's designers often describe their work as sculpting feelings rather than shapes. Every car must elicit a visceral response—whether it's awe, pride, or quiet admiration.

This is where the human element becomes irreplaceable. Despite all technological tools at their disposal, Aston Martin's designers still rely heavily on hand sketches, clay models, and intuition. Machines can calculate aerodynamics; only humans can understand beauty. This union of passion and precision gives every Aston Martin its distinctive heartbeat.

When a driver approaches an Aston Martin, the emotional experience begins long before ignition. The way light dances across its surfaces, the way the door opens with deliberate weight, the way the seat cradles the driver—it's all part of a symphony orchestrated by design. And when the car finally moves, that design translates emotion into motion.

Aston Martin's design isn't meant to impress—it's meant to **connect**. That connection is what transforms ownership into something deeply personal, something spiritual. It's why collectors treasure Aston Martins not as machines but as living works of art.

The next frontier for Aston Martin design will be defined by **sustainability, electrification, and digital craftsmanship**. Yet even as new technologies emerge, the company remains committed to its foundational ideals. The challenge ahead is to translate its design DNA into a silent, electric future without losing the brand's voice.

Future Aston Martins will likely feature **fluid digital surfaces**, minimalistic cabins, and adaptive aerodynamic systems—but they will still carry the essence of the DB lineage. Expect proportions that remain faithful to the brand's philosophy: long bonnets, short decks, and an unmistakable stance of readiness.

As Aston Martin steps deeper into an era of hybrid and electric performance, its design philosophy will ensure continuity through emotion. The roar of an engine may fade, but the **aesthetic pulse**—that unspoken energy of beauty and motion—will never be silenced.

To understand Aston Martin's design philosophy is to understand the soul of the brand itself. Every curve tells a story of passion, every line reflects discipline, and every proportion reveals purpose. Over more than a century, Aston Martin has turned automotive design into a poetic language—spoken fluently through steel, leather, and light.

The brand's enduring success lies in its ability to make people *feel*. In an age of automation, that emotional resonance has become its most precious commodity. Aston Martin's design remains a timeless dialogue between art and engineering, beauty and function, heart and hand.

For all its power and prestige, an Aston Martin is ultimately about **human connection**—the subtle magic that happens when beauty meets motion. It is not just designed to be seen; it is designed to be *felt*.

And that, more than anything, is the secret to Aston Martin's eternal allure—the philosophy that keeps its design truly **iconic through time**.

7: THE ASTON MARTIN DB9

When the **Aston Martin DB9** debuted in 2004, the automotive world took notice. It wasn't merely the introduction of another luxury grand tourer; it was a declaration — a statement that Aston Martin had re-entered the elite stage of automotive mastery with renewed vigor and purpose. For decades, the company had balanced precariously between heritage and innovation, often constrained by finances, changing ownerships, and the relentless pace of modern engineering. But the DB9 changed all that. It redefined what the words *"British luxury performance"* truly meant in the 21st century.

At the turn of the millennium, Aston Martin found itself standing at a crossroads. The company had a rich legacy built upon cars like the DB5, DB6, and V8 Vantage — machines that were as much about art and craftsmanship as they were about power and precision. Yet, the new millennium demanded something more: a modern interpretation of beauty, one that could rival the technological prowess and performance benchmarks set by Ferrari, Porsche, and Lamborghini. The DB9 became that bridge between past and future — a seamless blend of tradition and innovation, of handcrafted artistry and high-tech sophistication.

Aston Martin's CEO at the time, Dr. Ulrich Bez, envisioned a car that would elevate the brand into a new era of global recognition. This vision demanded not just a model refresh but a total reinvention — from the chassis up. And so, the DB9 was born, not as a continuation of the DB7, but as an entirely new creation designed to embody the brand's soul while propelling it into the future.

The DB9 was the first Aston Martin to be built on the **VH platform**, short for "Vertical/Horizontal" architecture — a modular aluminum structure designed to underpin multiple future models. This decision represented a monumental shift in Aston Martin's engineering philosophy. By adopting aluminum, the brand moved away from the heavy steel frames that had defined its earlier cars, resulting in a chassis that was **lighter, stronger, and more agile**.

The engineering team understood that for Aston Martin to compete on the world stage, the DB9 needed more than beauty — it needed brains beneath the bodywork. Aluminum panels were bonded and riveted using aerospace techniques, creating a structure that was both rigid and lightweight. The result was a car that handled with finesse and responded with precision — a true driver's machine wrapped in the luxury of a grand tourer.

Aesthetically, Henrik Fisker's design amplified the power of the new platform. Every curve and contour of the DB9's body was sculpted with aerodynamic efficiency in mind. The long bonnet, the sharply raked windscreen, and the smooth, muscular haunches conveyed both grace and aggression. The aluminum panels caught the light in ways that seemed to change the car's character with every angle — elegant at rest, predatory in motion. The result was timeless: a shape that, even decades later, continues to be hailed as one of the most beautiful automotive designs ever penned.

Underneath that flowing exterior, the DB9's engineering precision told another story — one of obsession. Each component, from the suspension to the exhaust routing, was optimized for performance and balance. Engineers worked tirelessly to ensure the weight distribution was nearly perfect, giving the DB9 a natural composure whether cruising at 30 mph through a city boulevard or carving a mountain pass at 120.

If the DB9's chassis was revolutionary, its design was transcendent. Fisker, known for his bold yet elegant approach to car design, captured Aston Martin's heritage in a body that was unmistakably modern. The **DB9's proportions** were near perfection — a long, sweeping hood, a compact cabin positioned toward the rear, and a low stance that exuded confidence. It wasn't flashy; it was effortlessly beautiful, the kind of design that didn't need to shout to command attention.

At the heart of this beauty was Aston Martin's **philosophy of proportion and purity**. Every line served a purpose. The front grille, larger than that of previous models, was assertive yet graceful — a visual signature that announced Aston Martin's return to prominence. The side strakes, subtle nods to earlier DB models, carried heritage forward, while the flowing roofline seamlessly tapered into a rear end sculpted for aerodynamic precision.

The DB9's design was a study in balance — **muscular without being overbearing**, elegant without appearing fragile. Its aluminum panels were hand-finished to perfection, with craftsmen at Aston Martin's Gaydon facility spending hours ensuring every contour reflected light flawlessly. Unlike the computer-generated surfaces of its rivals, the DB9's curves were organic, the result of artistic intuition guided by engineering precision.

Inside, the cabin reflected the same philosophy of handcrafted luxury. Every inch was a symphony of textures — buttery soft leather, genuine walnut veneer, cool aluminum switchgear. There was no plastic fakery here; every material was authentic, tactile, and pleasing to the senses. The DB9's interior wasn't designed for ostentation — it was designed for immersion. Sitting in the driver's seat felt less like entering a car and more like slipping into a tailored suit — one that fit perfectly, moved fluidly, and exuded quiet confidence.

Aston Martin's artisans in Gaydon treated each interior as a bespoke creation. The seats were hand-stitched, the dashboard meticulously assembled, and the attention to alignment and finishing bordered on obsessive. This craftsmanship wasn't about speed of production but about respect for tradition — a deeply ingrained belief that every Aston Martin should feel *alive* in the hands of its owner.

Powering the DB9 was the beating heart of Aston Martin's engineering excellence — a **5.9-liter naturally aspirated V12** engine, derived from a lineage that began in the DB7 Vantage. Producing **450 horsepower and 420 lb-ft of torque**, it wasn't merely an engine; it was a mechanical orchestra. Every press of the accelerator unleashed a deep, resonant symphony — a throaty growl that rose into a metallic wail as the revs climbed. This wasn't sound tuned for theatrics; it was the natural resonance of twelve cylinders working in perfect harmony.

The V12 engine gave the DB9 a personality unlike any other grand tourer. Smooth and refined during casual driving, it transformed into a ferocious beast when provoked. The car could sprint from 0 to 60 mph in just **4.6 seconds** and achieve a top speed near **190 mph**, but numbers alone don't capture its magic. The DB9 didn't just *go fast* — it *felt fast*. The engine's linear power delivery and intoxicating soundtrack created an emotional bond between car and driver that transcended performance figures.

Aston Martin's engineers ensured the V12 wasn't just powerful but perfectly integrated. Mounted low and far back in the chassis for optimal weight distribution, it gave the DB9 impeccable balance. Whether sweeping through high-speed corners or gliding along a coastal road, the car felt composed, predictable, and deeply connected to the driver's intentions.

The gearbox — a **six-speed automatic with paddle shifters** — provided versatility. In automatic mode, shifts were buttery smooth, perfect for grand touring. But in manual mode, the DB9 awakened. Each flick of the paddle delivered a satisfying surge of torque, accompanied by that unmistakable V12 growl echoing through the exhaust. It was mechanical poetry — the kind that stirred the soul of anyone lucky enough to command it.

Beyond its technical brilliance, the DB9 represented something deeper — the **rebirth of Aston Martin's spirit**. It wasn't just a car designed to compete; it was a statement of belief in human craftsmanship in an increasingly digital age. In a world rapidly surrendering to automation, the DB9 was still built by hand, each vehicle crafted by a small team of artisans who took immense pride in their work.

This approach was not just about tradition but about emotion. Every DB9 carried with it the fingerprints, the care, and the attention of those who built it. The connection between man and machine — between builder and driver — was tangible. When you sat behind the wheel, you didn't just experience engineering excellence; you experienced the culmination of passion, heritage, and human artistry.

Aston Martin's design and engineering teams believed in creating *living machines* — cars that weren't defined solely by their specifications but by how they made you feel. The DB9 embodied that philosophy to perfection. It was graceful yet potent, elegant yet wild, restrained yet emotional — the automotive equivalent of a symphony that could shift from a whisper to a roar in an instant.

The DB9 wasn't just a car for the road; it was a piece of rolling art, a symbol of what happens when human creativity meets technical mastery. Every drive felt like a ritual — an awakening of senses, a connection between driver and machine that bordered on the spiritual. This emotional resonance is what separated the DB9 from its competitors. While others pursued speed and numbers, Aston Martin pursued *feeling*. And that, ultimately, became its greatest triumph.

By the time the DB9 graced showrooms in 2004, it had already achieved something extraordinary: it reignited global fascination with Aston Martin. It wasn't simply a car — it was a renaissance. The DB9 proved that elegance and performance could coexist without compromise, that design and technology could harmonize without losing soul.

In an age of increasing digital detachment, the DB9 stood as a reminder that true beauty comes from craftsmanship, and true performance from passion. It set the foundation for every Aston Martin that followed — from the DB11 to the DBS Superleggera — each carrying echoes of the DB9's DNA in its lines, curves, and heartbeat.

By the time the **Aston Martin DB9** began rolling out of the company's Gaydon facility in 2004, it wasn't merely another luxury car launch—it was the beginning of a global phenomenon. The DB9 represented a rebirth, not just for Aston Martin, but for the entire British automotive industry. In an era dominated by the Italian flair of Ferrari, the German precision of Porsche, and the extravagant aggression of Lamborghini, Aston Martin emerged as the elegant disruptor—the refined alternative that blended passion with poise, technology with timelessness.

The DB9's presence was magnetic. When it debuted at the Frankfurt Motor Show, it wasn't greeted with polite applause—it was met with reverence. Journalists called it "the most beautiful car in the world." Buyers, collectors, and even rival designers took notice. The DB9 embodied the kind of beauty that transcended trends; it wasn't loud or ostentatious—it was *inevitable*. It looked like a car that *had to exist*, as if the very laws of design demanded it.

What made the DB9's impact so profound was its **ability to evoke emotion** in a way that few modern cars could. Every element of its design and engineering seemed orchestrated to connect with the driver on an emotional level. From the soft pulse of the V12 to the tactile engagement of its leather-wrapped controls, the DB9 invited not just admiration, but affection. It was a car that people didn't just want to own—they wanted to *believe in*.

This emotional connection became Aston Martin's calling card. The DB9 wasn't just competing in the luxury market—it was rewriting what that market meant. It showed that performance didn't have to scream to be heard, and that true luxury wasn't about excess, but about purity. In a world increasingly obsessed with speed records and digital gimmicks, the DB9 was refreshingly human—a masterpiece shaped not only by wind tunnels and CAD software, but by hands, eyes, and hearts.

Behind its graceful exterior, the DB9 was continuously evolving. Over its long production run, Aston Martin refined every mechanical and technological element, ensuring that the DB9 never stood still. The early models were already benchmarks, but by the time later editions such as the DB9 GT arrived, the car had matured into something extraordinary.

The aluminum **VH platform**—a marvel of modular design—proved its worth, allowing Aston Martin to continuously upgrade the DB9's capabilities without losing its essence. Engineers fine-tuned the suspension for sharper response and greater ride comfort, adjusted the transmission for quicker shifts, and gradually improved weight balance to perfect the driving dynamics. The result was a car that grew more sophisticated with every iteration, while still maintaining its unmistakable Aston Martin soul.

Perhaps the most notable evolution was in the **engine performance**. Initially producing 450 horsepower, the DB9's 5.9-liter V12 would eventually be tuned to deliver over 540 horsepower in the DB9 GT—an increase achieved without compromising refinement or character. Each upgrade brought smoother delivery, improved throttle response, and that unmistakable soundtrack—the deep, sonorous growl that could make a driver's heart race before the car even moved.

The DB9's adaptive damping system allowed drivers to choose between comfort and sport modes, seamlessly transforming the car from a long-distance cruiser into a performance machine. It wasn't just about speed—it was about *balance*. The DB9's ride quality became legendary, striking the perfect harmony between control and compliance. On highways, it floated gracefully; on winding roads, it danced with poise.

This duality—*grand tourer and sports car in one body*—became the foundation of Aston Martin's modern identity. Where rivals often forced buyers to choose between luxury and performance, the DB9 proved that both could coexist beautifully. It wasn't a compromise; it was a statement of mastery.

Step inside the DB9, and you entered a world defined by craftsmanship and calm. Unlike many supercars that intimidated their drivers, the DB9 welcomed them. The interior wasn't about digital overload or excessive flash; it was about refinement. The handcrafted cabin reminded you that this was not a machine mass-produced by robots, but a *creation* born from human artistry.

The **leather upholstery**, sourced from Bridge of Weir in Scotland, was of the highest grade—soft, supple, and hand-stitched with precision. The wood veneers were individually selected for grain consistency, then lacquered and polished to perfection. Aluminum controls, each cold to the touch, offered satisfying resistance when turned or pressed. Even the crystal key—known as the "Emotional Control

Unit"—was a poetic detail, symbolizing the connection between driver and machine.

Aston Martin's designers understood that the interior of a car wasn't merely a cabin; it was a *sanctuary*. The DB9's ergonomics were deliberately simple, uncluttered, and timeless. While competitors crammed their dashboards with touchscreens and flashing displays, the DB9 kept its focus pure: tactile control, mechanical honesty, and visual serenity. Everything had a purpose; nothing was superfluous.

Even at high speed, the DB9 felt tranquil. Its cabin isolation was near-silent, allowing the driver to glide across continents without fatigue. Yet, with a push of the throttle, serenity transformed into symphony—the V12's deep voice filling the cabin with life. This ability to merge comfort with exhilaration defined the DB9's luxury. It wasn't just about softness or quietness; it was about control—the confidence of knowing that beneath the surface elegance lurked unrestrained power.

Few cars achieve what the DB9 did in terms of **timeless design**. Even twenty years after its debut, it remains one of the most beautiful automobiles ever made—a design that feels both classic and contemporary. Henrik Fisker's original vision became the blueprint for an entire generation of Aston Martins. From the DB11 to the DBS Superleggera, echoes of the DB9's lines can still be traced: the long bonnet, the flowing waistline, the elegant rear taper.

But the DB9's influence extended beyond Aston Martin's own stable. Designers from rival marques often cited it as an example of perfect proportion—a car that balanced aggression and grace with effortless harmony. It became a teaching model in design schools, studied for its restraint, purity, and form. While others chased trends, the DB9 transcended them. Its design wasn't *of* its time—it was *beyond* time.

Aston Martin's approach to design with the DB9 was rooted in the belief that beauty should have integrity. Every crease, curve, and contour served both an aesthetic and aerodynamic purpose. The car wasn't adorned with unnecessary spoilers or vents; it was sculpted for performance through elegance. This philosophy became Aston Martin's visual DNA—recognizable at a glance, unmistakable in motion.

The DB9's physical presence was also matched by its emotional one. Seeing it parked in a driveway or gliding down a boulevard wasn't just seeing a car—it was witnessing a work of art in motion. The shape exuded confidence without arrogance, elegance without fragility. It spoke softly, but every line carried power.

And perhaps that is the greatest achievement of all—the DB9's ability to *move people*, not just physically, but emotionally. It reminded the world that cars could still be beautiful, not just functional. It reawakened the notion that driving could be a sensuous experience, an act of art and passion.

No Aston Martin chapter would be complete without discussing its cultural resonance. The DB9 carried the legacy of its cinematic forebears, most notably the DB5—immortalized by James Bond. While the DB9 itself never starred as 007's official car, its influence was felt across the franchise and beyond. It was the modern embodiment of what the DB5 represented: sophistication, danger, and allure.

Celebrities, collectors, and connoisseurs gravitated toward the DB9 not for status alone, but for what it *represented*. It was a symbol of taste—a choice made by those who valued refinement over ostentation. In Hollywood and high society alike, the DB9 became shorthand for understated success. You didn't buy a DB9 to impress others; you bought it to *express yourself*.

Magazines called it "the gentleman's supercar." It wasn't the loudest or the fastest, but it was arguably the most *complete*. It didn't shout for attention; it commanded it. The DB9 became a fixture at events like Pebble Beach, Goodwood, and the Monaco Grand Prix—gatherings where heritage and style meant as much as horsepower and speed.

Even within Aston Martin's own lineup, the DB9 held a unique place. It bridged the gap between the raw performance of the Vantage and the grandiose luxury of the Vanquish. It was the car that defined the brand's *sweet spot*—where performance met poetry, and design met desire.

When production of the DB9 finally ended in 2016, it marked the close of an extraordinary twelve-year chapter. Over that time, more than 16,500 units had been produced—each one meticulously hand-built, each one carrying its own soul. It wasn't just one of the longest-running Aston Martins ever made; it was one of the most loved.

But the DB9's story didn't end on the assembly line. Its legacy lives on—in the DB11, in the DBS Superleggera, and in every modern Aston Martin that continues to embody its philosophy. The lessons learned from the DB9's aluminum architecture, design integrity, and emotional appeal became the foundation of Aston Martin's future.

The DB9 also paved the way for Aston Martin's **redefinition as a global luxury icon**. It attracted a new generation of buyers from the United States, Europe, the Middle East, and Asia—markets that had previously looked elsewhere for high-end performance. The DB9 proved that British engineering could stand toe-to-toe with the best of Italy and Germany, not by imitation, but by innovation.

And perhaps that's the DB9's greatest legacy of all—it restored faith. Faith in design, in craftsmanship, in the idea that a car could still be *beautiful for beauty's sake*. It showed that passion could be engineered, and that driving could still stir the soul in an increasingly digital world.

The DB9 will forever stand as a turning point in Aston Martin's journey—a symbol of rebirth and resilience, of art meeting machinery, of man meeting masterpiece. It wasn't merely a car; it was a *promise kept*—that Aston Martin would never stop creating objects of desire, objects that move us in ways no numbers ever could.

Looking back now, the Aston Martin DB9 is not just remembered for what it was, but for what it inspired. It redefined expectations, reignited passion, and reestablished Aston Martin as the ultimate purveyor of **grand touring excellence**.

For collectors, the DB9 remains a touchstone—an object of enduring value and admiration. For designers, it remains a case study in restraint and proportion. And for drivers, it remains an experience that defies time—a machine that reminds us why we fell in love with cars in the first place.

Aston Martin's motto, "Power, Beauty, Soul," found its purest expression in the DB9. Power in its V12 heartbeat.

Beauty in its sculpted form. Soul in the craftsmanship and emotion it carried. Together, these elements created not just a car, but a legacy—a legacy that continues to define the very essence of what makes Aston Martin more than a brand, but a belief.

The DB9 was, and always will be, the car that taught the world a simple truth:

Perfection is not found in excess—it's found in elegance.

8: THE VANQUISH LEGACY

In the early dawn of the 21st century, Aston Martin found itself standing at a crossroads. The world was changing fast — technology, design, and consumer expectations were all evolving at an unprecedented pace. For a brand steeped in old-world craftsmanship and traditional British elegance, this was a defining moment. The company needed a new hero — a car that could not only uphold the spirit of its illustrious past but also propel it into a bold, modern future. That car would be the **Aston Martin Vanquish**, a name that carried both promise and power, and a machine that would go on to redefine what it meant to be a luxury sports car.

When the covers came off at the **2001 Geneva Motor Show**, the crowd was mesmerized. Under the dazzling lights stood a machine that radiated presence — muscular yet poised, sophisticated yet aggressive. The **Vanquish** was not merely a new model; it was a **declaration of intent** from Aston Martin. This was a car designed to prove that British engineering could still stand tall against Italian finesse and German precision.

Its purpose was clear — to reassert Aston Martin's dominance in the world of high-performance grand tourers, a segment it had long defined with models like the **DB4**, **DB5**, and **V8 Vantage**. Yet, this time, the brand wanted more than admiration; it wanted respect from a new generation of enthusiasts and collectors.

The man entrusted with shaping this new flagship was **Ian Callum**, one of the most respected automotive designers of his era. Callum's challenge was monumental: he had to create a car that felt unmistakably Aston Martin, yet distinctively modern. It couldn't just echo the brand's past glories — it had to define the aesthetic of its future.

The result was nothing short of breathtaking. The Vanquish embodied **the perfect tension between elegance and aggression**. Its long hood and tightly drawn proportions exuded timeless Aston Martin grace, while its pronounced haunches and low, athletic stance conveyed explosive potential. Every surface seemed to have been sculpted by airflow itself, from the smooth curvature of the fenders to the subtle flick of the rear spoiler integrated seamlessly into the body.

Callum described his design philosophy for the Vanquish as *"muscular elegance."* He wanted the car to look alive, like a wild animal ready to leap at any moment. The front fascia, with its signature Aston Martin grille, carried a sharper, more assertive character than any model before it. The headlights, sleek and piercing, gave the car a predatory expression — one that suggested power lurking beneath the surface.

The car's **carbon-fiber composite body panels** were not merely aesthetic choices but functional innovations. This was one of the first Aston Martins to integrate such advanced materials into its structure, setting a precedent for future models. The blend of aluminum and carbon fiber offered both lightness and strength, giving the Vanquish an agility that belied its grand touring nature.

At the heart of the Vanquish was its **5.9-liter naturally aspirated V12 engine**, an engineering masterpiece that embodied the soul of Aston Martin. Producing an impressive **460 horsepower and 400 lb-ft of torque**, this engine gave the car ferocious performance without sacrificing the smooth refinement expected of a luxury grand tourer.

The V12's roar was something to behold — a symphony of mechanical precision that crescendoed into a spine-tingling growl as the tachometer climbed. It was a sound that could stir emotions before the car even moved. For many enthusiasts, the Vanquish wasn't simply a car; it was a **sensory experience**. From the vibration through the steering wheel to the deep resonance of the exhaust, every moment behind the wheel was visceral.

Acceleration from 0–60 mph took a mere **4.1 seconds**, and the car could surge effortlessly past **200 mph**. But these figures only told part of the story. The real magic of the Vanquish lay in how it delivered its performance — smooth, composed, and utterly confident. It wasn't just fast; it was dignified in its speed, a gentleman's express that could duel with the world's best supercars yet carry itself with aristocratic poise.

Behind the beauty of the Vanquish was a revolution in engineering philosophy. Aston Martin's engineers knew that for the car to compete at the highest levels, it had to shed the heavy, traditional construction methods of the past. Enter the **VH platform** — short for **Vertical/Horizontal architecture** — a new modular approach that would underpin many of the brand's future models.

The Vanquish's chassis was a **hybrid of aluminum and carbon fiber**, bonded together to create a lightweight yet immensely strong structure. This construction technique marked a turning point for Aston Martin. Not only did it enhance rigidity and safety, but it also allowed for sharper handling and improved balance.

Complementing the advanced chassis was a **sophisticated double-wishbone suspension system** that ensured the perfect balance between comfort and control. Whether carving through mountain passes or gliding along the motorway, the Vanquish delivered an experience that was both exhilarating and effortless.

But perhaps the most debated feature of the Vanquish was its **electrohydraulic manual transmission**, known as the **"Touchtronic"** system. Operated via paddle shifters mounted behind the steering wheel, this semi-automatic gearbox offered drivers the option of manual control without a traditional clutch pedal. It was cutting-edge for its time — inspired by Formula 1 technology — but it divided opinion among purists. Some loved its immediacy and futuristic nature; others longed for the tactile engagement of a true manual.

Nevertheless, the Touchtronic system represented Aston Martin's willingness to innovate and push boundaries — an attitude that had not always been associated with the brand. The Vanquish, in many ways, symbolized **the fusion of tradition and technology**, of hand craftsmanship meeting digital precision.

While the Vanquish was a marvel of engineering, it was inside the cabin that Aston Martin's legendary craftsmanship truly shone. Each interior was meticulously hand-assembled by artisans at the company's historic **Newport Pagnell facility**, a site that had become synonymous with bespoke British luxury.

Step inside a Vanquish, and you were greeted by an environment of pure elegance. **Hand-stitched leather** adorned nearly every surface, with contrasting threads that highlighted the precision of the craftsmanship. **Polished aluminum switchgear**, **real wood veneers**, and **carbon-fiber trim** blended seamlessly, creating an atmosphere that was both classic and contemporary.

Every control had a satisfying tactile quality — the way the door latch clicked shut, the weighted feel of the gear paddles, the gentle resistance of the leather-wrapped steering wheel. This was not a mass-produced interior; it was a **work of art** crafted for the discerning driver.

Even the layout of the dashboard and center console was a study in balance. Unlike the cluttered interiors of some of its competitors, the Vanquish embraced simplicity and proportion. The **central starter button**, elegantly encased in glass, was a theatrical touch — a small ceremony that transformed every start-up into a moment of drama.

Driving the Vanquish was an experience unlike any other. It wasn't a car that demanded to be tamed; it **invited collaboration**. The V12 engine responded to the driver's inputs with precision, while the car's balanced chassis inspired absolute confidence. Even at high speeds, the Vanquish remained composed, its aerodynamics and suspension working in perfect harmony.

The car's steering was direct yet fluid, providing a clear sense of connection to the road. Its brakes — massive ventilated discs with advanced ABS — delivered stopping power that matched the car's immense performance. Whether navigating tight bends or cruising down a coastal highway, the Vanquish was both **a beast and a ballet dancer** — ferocious in spirit, graceful in motion.

Yet, what truly set the Vanquish apart was how it made its driver feel. It wasn't just about speed or handling — it was about **presence**. Sitting behind that sculpted hood, with the world unfolding ahead, you felt part of something grander — a lineage of craftsmanship, power, and art.

For many owners, the Vanquish was not a mere possession; it was an **emotional statement** — a declaration of taste, ambition, and appreciation for mechanical beauty.

The Vanquish's allure extended far beyond the automotive world. Its cinematic debut as **James Bond's car in "Die Another Day" (2002)** propelled it into pop culture immortality. Outfitted with all the high-tech gadgetry befitting 007, the Vanquish — or the "Vanish," as Q humorously dubbed it — became an instant symbol of sophistication and danger.

For Aston Martin, this was more than a marketing coup; it was a cultural moment. The car's appearance on the silver screen introduced the Vanquish to millions who might never have stepped into a showroom. It rekindled the romance between Aston Martin and Bond, a relationship that had begun with the DB5 in *Goldfinger* and would continue for decades.

In that moment, the Vanquish was no longer just a car — it was **an icon of British excellence**, embodying everything that made the brand special: beauty, power, and mystery.

By the time production ended in 2007, the Aston Martin Vanquish had achieved something few cars ever do — it had become **a legend in its own lifetime**. Its influence on later models, such as the DBS, DB9, and the second-generation Vanquish, was profound.

But beyond its technical legacy, the car left a deeper mark — it redefined what Aston Martin stood for in the modern age. The Vanquish proved that the brand could evolve without losing its essence. It married **heritage with innovation**, **luxury with performance**, and **emotion with engineering** in a way that few competitors could match.

Even today, the Vanquish commands admiration not just as a collector's item but as a symbol of an era when Aston Martin rediscovered its soul. It wasn't just a machine; it was a manifesto — one written in aluminum, leather, and fire.

The Aston Martin Vanquish wasn't just a car—it was the declaration of a new epoch. When the last echoes of the V12's roar reverberated across the halls of the Geneva Motor Show in 2001, automotive journalists, enthusiasts, and rivals all recognized one truth: Aston Martin had entered a bold, uncharted chapter. The Vanquish had not only revived the brand's spirit—it had redefined it.

For years, Aston Martin had been synonymous with grace and heritage, but the Vanquish turned that tradition into momentum. Its aluminum and carbon-fiber chassis represented not just progress but vision—an understanding that the future of performance would rest upon the harmony of lightness and strength. This was not a car built merely to impress; it was engineered to transform how drivers connected with speed, emotion, and art.

The Vanquish also marked Aston Martin's transition from handcrafted nostalgia to modern mastery. It was a mechanical symphony of the past and the future—where analog passion met digital precision. The technological leap it made was a signal to the automotive world: **Aston Martin was no longer chasing competitors—it was defining its own horizon.**

At its core, the Vanquish carried forward Aston Martin's grand touring DNA—a car designed for both power and poise, for acceleration and serenity. Yet what made it special was how it redefined what a grand tourer could be.

The V12 wasn't merely an engine; it was an orchestra of twelve cylinders, each tuned to perfection, producing a sound that could send chills down a driver's spine. It wasn't loud—it was operatic. Underneath that thunder, however, lay elegance. The Vanquish wasn't raw aggression; it was sophistication refined through speed.

On winding alpine passes, the Vanquish danced with balance—its chassis nimble, its steering communicative, its power smooth and unrelenting. On the autobahn, it became a missile cloaked in luxury, capable of traveling at immense speed while cocooning its occupants in tranquility. And in city streets, it carried itself with the dignity of a tailored suit—commanding attention without needing to shout.

Aston Martin didn't just want to build a supercar; it wanted to build an experience—one that captured emotion as much as acceleration. That duality became the Vanquish's identity, and ultimately, its legend.

Designing the Vanquish was no small feat. Ian Callum, Aston Martin's visionary design director, approached the car as both a sculpture and a statement. Every curve, vent, and contour was crafted to convey power through beauty. The design didn't just age gracefully—it defied time.

The Vanquish's proportions were near-perfect: long hood, short rear deck, and a cabin pushed toward the back—classic grand tourer aesthetics reborn with modern aggression. The way the fenders flared subtly over the wheels gave it a stance of latent energy, as if it could surge forward at any moment. Its eyes—sharp, focused headlamps—gave the impression of intelligence and intensity.

What few saw, however, was how meticulously the design blended form with function. The carbon fiber front splitter, the sculpted rear diffuser, the smooth underbody—all contributed to aerodynamic stability. The Vanquish wasn't only beautiful; it was intelligent beauty.

Step inside, and you entered a world that felt more like a gentleman's club than a cockpit. The leather was hand-stitched, the switches milled from aluminum, and the seats contoured to perfection. The dashboard flowed like the instrument panel of a private jet—luxurious yet purposeful. Every touch reminded the driver that this was not mass production—it was a masterpiece.

No discussion of the Vanquish would be complete without mentioning its most iconic appearance—**as the star car of James Bond in *Die Another Day* (2002)**. Dubbed the "Vanish" due to its cloaking technology in the film, it cemented the car's image as the ultimate symbol of elegance and power.

The Bond association wasn't new for Aston Martin—after all, the DB5 had become legendary in *Goldfinger*—but the Vanquish gave the relationship a modern twist. Sleek, technologically advanced, and unapologetically British, it was the perfect extension of Bond's persona.

The exposure brought the Vanquish into the cultural limelight. It wasn't just admired by car enthusiasts anymore—it became a dream machine recognized by millions worldwide. To own a Vanquish was to possess a piece of cinematic history, a car that symbolized both style and danger, precision and passion.

A decade later, Aston Martin resurrected the legend. In 2012, the **second-generation Vanquish** arrived—a car that built upon the legacy of its predecessor while embracing the technologies of a new era.

This Vanquish featured a re-engineered version of Aston Martin's V12, now producing over 560 horsepower, coupled with a new carbon-fiber body and enhanced aerodynamics. It was faster, lighter, and sharper than ever before. Yet despite its modern updates, it retained the emotional core that defined the original—a blend of sophistication and savagery.

Inside, the new Vanquish introduced digital displays, a revamped infotainment system, and refined ergonomics. The craftsmanship was still unmistakably Aston Martin—every stitch, every contour exuded bespoke artistry. It was a car that embraced modern luxury without abandoning heritage.

When the Vanquish S arrived in 2017, it pushed boundaries even further—600 horsepower, sharper handling, and the last naturally aspirated V12 grand tourer from Aston Martin. It was the end of an era—a farewell note written in pure gasoline and passion.

The Vanquish's influence didn't end with its production. It became the template for every Aston Martin that followed—the DBS, DB11, and even the latest Valour and Vanquish Vision Concept all carried echoes of its DNA. Its silhouette, its sound, its soul—they were reborn, reinterpreted, but never forgotten.

Collectors today still regard the early 2000s Vanquish as a defining moment in Aston Martin's history—a car that bridged the analog past with the digital future. It represented courage in design and confidence in execution. More than any other model, it reminded the world what Aston Martin stood for: **power, beauty, and soul.**

Even now, decades later, the Vanquish continues to stir emotion. Its legacy lives not just in showrooms or garages, but in the hearts of those who understand that true greatness is timeless.

It was never just a car—it was a declaration, a symphony of speed and elegance, a moment in time where art met engineering, and together they created something immortal.

In the world of automobiles, very few machines transcend their mechanical purpose. The Aston Martin Vanquish did. It was a car that redefined what it meant to move—physically, emotionally, and spiritually.

For Aston Martin, the Vanquish was more than a model line; it was a vision realized. It proved that performance could coexist with grace, and that luxury could live in harmony with ferocity. It was the bridge between the golden past and the electrified future.

Even as Aston Martin moves into an era of hybrid and electric innovation, the echo of the Vanquish's V12 lingers—a reminder of a time when driving was pure, passionate, and profoundly human.

The Vanquish will forever remain what its name promises: the conqueror—the car that vanquished doubt, redefined beauty, and secured its throne in the pantheon of automotive immortality.

9: PERFORMANCE AND ENGINEERING EXCELLENCE

There are few names in the world of motoring that evoke the same sense of artistry, passion, and precision as **Aston Martin**. Each car that rolls out of its workshop is not merely an automobile — it is a fusion of **mechanical poetry and human emotion**, engineered with the same devotion that a maestro conducts an orchestra. The essence of Aston Martin's engineering has always been its ability to **seamlessly merge raw power with refined beauty**, ensuring that every drive is not just fast, but deeply felt.

To understand Aston Martin's mastery, one must look beyond numbers and performance statistics. The brand's engineering is rooted in philosophy — a commitment to balance, proportion, and elegance. While other manufacturers chase acceleration figures and lap times, Aston Martin pursues a deeper ideal: *the perfect harmony between man, machine, and motion.* This harmony is what defines **Performance and Engineering Excellence**, the very DNA that courses through the veins of every Aston Martin produced.

If there is a single element that captures the essence of Aston Martin's engineering prowess, it is the **V12 engine** — a masterpiece of mechanical engineering that has become the beating heart of the brand for decades. This engine is not just a configuration of pistons and valves; it is a symbol of craftsmanship, heritage, and relentless pursuit of perfection.

Aston Martin's V12 is an orchestra of twelve cylinders, each one tuned to perfection, producing a sound that has become the brand's signature symphony. It is the sound of tradition meeting technology — deep, resonant, and unmistakably British. Unlike the harsh metallic scream of Italian supercars or the turbocharged hiss of German engineering, the Aston Martin V12 sings in a tone of power refined by grace.

The engine's **60-degree configuration** ensures balance and smoothness, allowing the car to glide effortlessly, even at extreme speeds. This geometry minimizes vibration and mechanical harshness, creating a sensation of fluid motion. The result is a car that does not just accelerate — it **surges forward with elegance**, maintaining composure and grace at every revolution.

From the **DB7 Vantage** to the **Vanquish and DBS**, the V12 has evolved but never lost its character. The early 5.9-liter versions produced around 450 to 510 horsepower, while later iterations, such as in the DBS Superleggera, would deliver a staggering 715 horsepower. But numbers only tell half the story. The real magic of the V12 lies in its **throttle response and torque delivery** — instant, smooth, and linear. Power builds with the precision of a rising crescendo, giving the driver complete command without intimidation.

When you press the starter button in an Aston Martin V12, you don't just start an engine — you awaken a legacy. The low, rhythmic idle feels alive, as if the car itself is breathing, waiting for your next move. Then, with a slight touch on the accelerator, the symphony begins — deep and soulful, echoing through your chest as much as your ears. This is not just an engine; it is an experience, a living connection between man and machine.

Engineering excellence at Aston Martin is not defined solely by horsepower; it is defined by **how that power is delivered**. Every component, from the chassis to the suspension geometry, is designed to translate energy into emotion. The company's engineers operate with an artist's precision, where metal becomes muscle and speed becomes sculpture.

Central to this philosophy is Aston Martin's use of **lightweight materials and structural innovation**. The introduction of bonded aluminum architecture marked a revolution in the brand's engineering language. Unlike traditional welding, the bonding process uses aerospace-grade adhesives and rivets to fuse panels together. The result? A chassis that is not only lighter but also **incredibly rigid**, providing both safety and precision handling.

This structural integrity allows for a lower center of gravity and more accurate steering response. Combined with the brand's finely tuned suspension systems, it gives the driver an unparalleled sense of connection. Whether cornering at high speed or cruising along a coastal highway, an Aston Martin always feels balanced — poised yet alive, stable yet exhilarating.

Aston Martin's engineers understand that **performance is not aggression; it is grace under pressure**. This philosophy is evident in how the car transitions through every curve. The suspension absorbs imperfections with elegance, while the steering remains communicative, delivering feedback that feels organic, never artificial. In an age where digital systems often insulate the driver from the road, Aston Martin insists on preserving the tactile purity of driving.

Every Aston Martin is more than a sum of its mechanical parts — it is an emotional instrument, tuned for the senses. The process of performance tuning at Aston Martin is as much art as it is science. Engineers spend thousands of hours refining the smallest details to ensure that every moment behind the wheel feels **instinctive and alive.**

The **suspension system** lies at the core of this tuning process. Aston Martin's **double-wishbone design** at both the front and rear allows for unparalleled control of wheel motion, ensuring that the tires remain in perfect contact with the road surface. This translates into predictable handling, sharp cornering, and an uncanny sense of stability — even at the limits of traction.

The introduction of **adaptive damping systems** further enhanced this balance. Sensors constantly monitor road conditions and driver inputs, adjusting the stiffness of the dampers in real time. The car seamlessly transitions from a supple grand tourer to a razor-sharp sports machine, adapting to your mood and the road ahead.

Yet even with such technology, the soul of Aston Martin's handling remains mechanical and analog at heart. The engineers emphasize feel — the subtle communication between the tires, the steering wheel, and the driver's fingertips. It's this feedback loop that transforms driving from a task into an experience, one that engages the senses and emotions alike.

Performance tuning at Aston Martin extends far beyond the engine and suspension — it encompasses the entire driving ecosystem. From aerodynamics to braking systems, every element works in harmony to deliver an experience that feels effortless, composed, and exhilarating.

Take, for instance, the **carbon-ceramic braking systems** found in Aston Martin's high-performance models. These brakes, derived from Formula One technology, offer superior stopping power while remaining incredibly lightweight. Unlike traditional steel brakes, carbon-ceramic rotors can withstand extreme temperatures without fading, ensuring consistent performance lap after lap. The pedal feel is progressive — firm, communicative, and confidence-inspiring.

Aerodynamics, too, play a crucial role. Aston Martin's design philosophy is not about chasing downforce at the expense of elegance; it's about **sculpting airflow with beauty and intent**. Every curve, vent, and contour serves both form and function. The underbody diffusers channel air efficiently, while the subtle rear spoilers provide stability without disrupting the car's aesthetic purity.

This is where Aston Martin stands apart from its competitors. While others may adopt aggressive splitters and towering wings, Aston Martin maintains restraint. Its cars are **athletic sculptures**, designed to cut through the air gracefully rather than forcefully. The result is a car that feels planted at high speeds while maintaining its timeless aesthetic appeal.

If there is one element that defines the Aston Martin experience as much as performance or design, it is **sound**. The exhaust note of an Aston Martin is not accidental — it is composed, engineered, and tuned with the same care as a symphony.

Aston Martin engineers treat sound as an integral part of performance, not a byproduct. Every pipe, chamber, and valve is designed to create a **sonic signature** that captures both aggression and refinement. At idle, the note is deep and restrained; at full throttle, it transforms into a soaring crescendo that seems to tear the air itself apart. It is a voice that demands attention — not loud for the sake of volume, but **resonant with identity and purpose**.

The engineers' goal is to make the car sound alive. Even the acoustics of the cabin are tuned so that the driver can feel the vibration of the V12 through the steering wheel and the seat, creating an intimate, visceral connection. This is mechanical theatre at its finest — every rev, every downshift, every burble designed to evoke emotion.

Behind every Aston Martin lies not just technology, but **human craftsmanship**. Engineers, designers, and test drivers collaborate with a shared mission: to create machines that evoke emotion as much as admiration. The final tuning phase of every Aston Martin involves test drivers who rely as much on instinct as on data. They evaluate how the car feels, how it responds, and how it *communicates* with its driver.

This blend of empirical precision and emotional intelligence is what gives Aston Martin cars their unique personality. They are not cold, clinical machines built solely for performance metrics — they are living, breathing entities that express the vision and soul of their creators.

Aston Martin's commitment to this human touch is what separates it from others. In a world dominated by algorithms and automation, Aston Martin remains **an artisan of**

performance, where emotion and engineering coexist in perfect balance.

At the very core of the Vanquish's allure lies its balance — that near-perfect dance between power and poise. The engineers in Gaydon didn't want the car to feel like a brute force machine. They wanted finesse. The Vanquish's dynamics were honed to deliver confidence, allowing drivers to explore its immense performance without fear or fatigue. Every control input, from the steering to the brake pedal, was calibrated to speak clearly to the driver.

The chassis, a lightweight bonded aluminum structure complemented by a carbon fiber body, represented Aston Martin's commitment to modern materials science. This blend wasn't just for aesthetics — it was a calculated move toward rigidity, responsiveness, and safety. The result was a structure that felt both solid and alive, communicating every ripple of the road through the steering wheel, yet isolating occupants from harshness. It was a symphony of sensations — tactile, mechanical, and utterly human.

On winding mountain passes or racetrack straights, the Vanquish displayed a character that was both athletic and aristocratic. The steering's weight built naturally as cornering loads increased, offering a transparency that many modern systems still struggle to replicate. The hydraulic setup — chosen over electric assist for its purity of feel — became a hallmark of the car's communicative nature. It wasn't just steering; it was storytelling through motion.

Much of what made the Vanquish special lay hidden beneath its sculpted skin. The double-wishbone suspension, forged aluminum components, and adaptive damping system worked together in perfect harmony. Drivers could tailor the car's character — from a refined grand tourer to a sharp-edged sports machine — with the press of a button. Yet, no matter the setting, the Vanquish never lost its composure. It was controlled aggression, a predator at rest until provoked.

Aston's engineers paid extraordinary attention to suspension geometry. They understood that even a millimeter of deviation could alter the car's balance and feedback. The Vanquish wasn't meant to chase numbers; it was meant to *feel right*. Its grip levels were astonishing, yet it never felt clinical. There was an organic elasticity to its movements — the kind of grace that comes only from mechanical purity and decades of craftsmanship.

During wet British mornings or high-speed runs on continental highways, the stability control system was an invisible guardian. It intervened subtly, never breaking the car's rhythm or spirit. To enthusiasts, it felt as if the car had a sixth sense — a guardian angel fine-tuned to perfection. That's what separated the Vanquish from its rivals. It didn't dominate the road; it harmonized with it.

The Vanquish represented a critical turning point for Aston Martin — the moment when the brand embraced technology without sacrificing tradition. The car's evolution was a masterclass in subtle sophistication. The use of carbon fiber was more than a nod to motorsport; it was an assertion that lightweight innovation could coexist with handcrafted luxury.

The advanced Touchtronic gearbox, derived from racing technology, allowed seamless transitions between serenity and fury. Each upshift came with a crisp bark from the exhaust, each downshift accompanied by a mechanical symphony that stirred the soul. Aston Martin understood that performance wasn't just about acceleration — it was about *emotion through engineering*. Every gear change, every sound, every vibration was intentional. It created a bond, not just an experience.

Modern iterations of the Vanquish continued this philosophy. Improved aerodynamics, adaptive damping, and electronic torque vectoring made the car smarter and sharper, yet never disconnected. The relationship between man and machine remained sacred — unbroken by digital barriers. For Aston Martin, technology was not about replacing emotion; it was about enhancing it.

Perhaps no aspect of the Vanquish is as unforgettable as its voice. The 6.0-liter V12 didn't merely produce power — it composed music. From a low, rolling thunder at idle to a soaring metallic crescendo past 6,500 rpm, the sound was engineered to stir something primal. Engineers spent months tuning the intake resonance and exhaust note, treating acoustics as an art form. The result? A soundscape that felt alive, commanding, and unmistakably Aston.

On long drives, that soundtrack became part of the car's identity. Passersby could recognize a Vanquish before seeing it, drawn by its distinctive growl echoing across valleys or tunnels. This wasn't mere noise — it was storytelling in motion, a declaration of presence, of passion, of performance elevated to poetry.

Inside the cabin, the acoustics were tuned with equal care. The engine's voice was filtered, not muted, ensuring the driver always felt connected. It was this fusion of emotion and engineering that made the Vanquish not just fast, but *alive*.

As the Vanquish gained international acclaim, it became more than a car — it became a symbol. For many, it was the ultimate Aston Martin: a machine that captured the essence of British engineering wrapped in timeless beauty. Each line on its body was both functional and poetic, every surface designed to catch light in a way that made the car seem perpetually in motion, even at rest.

But perhaps its greatest triumph was how it connected the past and future of Aston Martin. It drew inspiration from icons like the DB5 and the V8 Vantage, yet looked ahead toward a new era of carbon technology and aerodynamic sophistication. It stood as proof that modern engineering and classic soul could coexist — not as opposites, but as complements.

From the sweeping bonnet to the muscular haunches, the Vanquish was a masterclass in proportion. Its design wasn't loud; it was composed. Understated power, quiet confidence — qualities that defined Aston Martin's philosophy for over a century.

Every great car leaves a mark, not just on the road but on the collective imagination of those who love driving. The Vanquish was one of those rare machines. It inspired designers, influenced rivals, and set a benchmark for how performance could coexist with grace. Even as the world shifts toward electrification, the spirit of the Vanquish remains relevant — a symbol of mechanical purity and human passion.

Its successors, like the DBS Superleggera and future electric GTs, carry its DNA forward — proof that while technology evolves, the essence of driving need not disappear. The Vanquish was never about being the fastest car in the world; it was about being the most *complete*. The one that could make your heart race not because of fear, but because of *feeling*.

To drive a Vanquish was to experience what Aston Martin called *"Power, Beauty, Soul"* — a mantra that transcends marketing and becomes truth the moment the V12 awakens. It was, and still is, an automotive masterpiece that doesn't just move you forward, but moves you *deeply*.

In the grand narrative of automotive evolution, few names evoke such reverence as the Vanquish. It represents more than performance metrics or luxury craftsmanship. It stands for the relentless pursuit of *perfection through passion*. For those who've driven it, it's not a machine — it's an experience etched into memory.

In the silence that follows the roar of its V12, one truth remains:

Great cars are not built merely to perform. They are built to inspire.

And the Aston Martin Vanquish will continue to inspire — generation after generation — as the car that redefined what it means to be truly *alive behind the wheel.*

10: THE ONE-77

In the pantheon of hypercars, few vehicles carry the weight of expectation and the aura of perfection quite like the Aston Martin One-77. Introduced in 2008, this hypercar was conceived not merely as another model in the storied lineup of Aston Martin but as the culmination of decades of engineering, design, and luxury craftsmanship. The One-77 was the embodiment of ambition, a car that would push the boundaries of what a road-going hypercar could achieve while remaining true to the very soul of Aston Martin: elegance, precision, and unrelenting performance.

From its conception, the One-77 was intended to be the ultimate statement of exclusivity. Every decision—mechanical, aesthetic, and technological—was aimed at creating a machine that would be celebrated as much for its engineering prowess as for its breathtaking beauty. With a production run strictly limited to seventy-seven units, the car's rarity instantly positioned it as a coveted masterpiece for collectors and automotive enthusiasts alike. Yet the One-77 was far more than a status symbol; it was a laboratory for innovation, a proving ground for Aston Martin's boldest ideas, and a declaration that the brand was capable of engineering a car that could rival the world's finest hypercars.

The story of the One-77 began in the early 2000s, during a period of transition for Aston Martin. The company, then under the ownership of Ford, was exploring ways to elevate its flagship offerings beyond traditional luxury grand tourers. The idea of a limited-production hypercar emerged: a vehicle that would combine extreme performance with Aston Martin's timeless design language and handcrafted luxury interiors. The number seventy-seven itself, chosen as the production limit, was not arbitrary—it symbolized rarity, prestige, and the brand's willingness to create something extraordinary.

The project officially began in 2005, with engineers, designers, and craftspeople collaborating to bring the One-77 to life. Aston Martin's design chief at the time, responsible for translating vision into tangible form, worked tirelessly alongside engineers to ensure that the car would not only perform at the highest levels but also remain instantly recognizable as an Aston Martin. Every line, curve, and detail was meticulously considered. From the flowing roofline to the aggressive front fascia, the One-77 was designed to be both aerodynamically efficient and visually breathtaking.

Engineering teams faced the challenge of developing a vehicle that could house immense power without compromising handling, stability, or safety. To achieve this, Aston Martin turned to advanced materials and cutting-edge design philosophies, ensuring that the One-77 would be as light, rigid, and responsive as possible. The result was a car that blended traditional craftsmanship with modern engineering ingenuity, creating a hypercar that could not only rival its contemporaries but, in many ways, surpass them.

At the heart of the One-77 is its **7.3-liter V12 engine**, a naturally aspirated marvel co-developed with Ford and Yamaha. This engine produces a staggering **750 horsepower** and 553 lb-ft of torque, allowing the One-77 to accelerate from zero to sixty miles per hour in just a few seconds. The power delivery is linear yet explosive, creating a driving experience that is as thrilling as it is precise. Unlike many modern hypercars that rely on forced induction, Aston Martin opted for a naturally aspirated engine, ensuring that the vehicle retained the high-revving character and unique auditory experience that defines a true V12. The exhaust note itself is a symphony of engineering—a throaty, melodic growl that evolves with engine speed, creating a visceral connection between car and driver.

Complementing the formidable engine is the One-77's **carbon fiber monocoque chassis**, which provides exceptional torsional rigidity while maintaining a remarkably low weight. This chassis forms the foundation for the car's dynamics, allowing the suspension to operate at peak efficiency and giving drivers confidence in every corner. The car's carbon fiber construction extends to body panels, front subframes, and even select interior elements, reducing mass without compromising strength or safety.

The suspension system, developed in collaboration with Formula One engineers, features **double-wishbone geometry** at both the front and rear. Coupled with adaptive dampers and a rear-mounted transaxle, this setup provides a perfect balance of agility and comfort. Whether navigating tight corners on a racetrack or cruising along scenic highways, the One-77 demonstrates a poise that is rare in a car of its power and speed.

Engineers also incorporated cutting-edge aerodynamics into the chassis design, ensuring optimal airflow and downforce at extreme velocities.

Beyond its mechanical sophistication, the One-77 represents a technological tour de force. Advanced electronic systems such as **traction control** and **electronic stability programs** work seamlessly with the mechanical components to maximize grip, stability, and driver confidence. These systems are calibrated to respond dynamically, allowing for a level of precision and control that ensures performance without compromising safety.

The braking system employs **carbon-ceramic discs**, designed to withstand repeated high-speed braking without fade. This ensures that even at speeds approaching 220 mph, the car can decelerate efficiently, offering both reassurance and performance. The braking system is paired with high-performance calipers and advanced electronic brake distribution, ensuring optimal stopping power under all conditions.

Adaptive aerodynamics further distinguish the One-77. A deployable **rear spoiler** adjusts automatically to speed and driving conditions, providing additional downforce at high speeds while retracting at lower speeds to reduce drag. This system works in harmony with front diffusers and underbody channels to create a car that feels planted yet agile, demonstrating Aston Martin's mastery of aerodynamic design in conjunction with traditional craftsmanship.

Even inside the cabin, the One-77 pushes boundaries. A bespoke **touch-screen interface** allows the driver to manage navigation, climate control, and audio systems with ease. Every element is designed to integrate seamlessly with the car's dynamic performance, ensuring that technology enhances rather than distracts from the driving experience.

Aston Martin's design ethos has always emphasized elegance and performance, and the One-77 represents the pinnacle of this philosophy. Every curve, every line, and every contour of the vehicle is meticulously sculpted to convey motion even when stationary. The car features a **long, sleek bonnet**, short overhangs, and a low-slung profile, creating a sense of forward momentum and aggression.

The front fascia features a wide, signature grille flanked by angular headlights, establishing a commanding presence on the road. The rear is equally dynamic, with muscular haunches, quad exhaust outlets, and a tapering tail that emphasizes both stability and style. Attention to detail extends to the interior, where luxurious leather, exotic woods, and brushed aluminum are combined in a tactile, aesthetically harmonious environment. Bucket-style seats provide lateral support during spirited driving while maintaining comfort for long journeys, exemplifying Aston Martin's philosophy of blending luxury with performance.

Each One-77 is highly **customizable**, allowing owners to choose exterior colors, interior materials, stitching, and trim. This bespoke approach ensures that no two cars are alike, emphasizing the exclusivity and personal connection between the car and its owner.

The Aston Martin One-77 is as much a statement of exclusivity as it is a hypercar. With only 77 **units produced**, each One-77 is a rare gem in the automotive world. This limited production run not only reinforced the car's status as a collector's item but also reflected the painstaking hand-built craftsmanship that goes into every vehicle. Aston Martin deliberately chose this number to emphasize scarcity, ensuring that owning a One-77 would remain an achievement reserved for a privileged few.

Each car was individually **hand-assembled** by a team of master engineers and craftsmen, a process that took several months per vehicle. No detail was too small to be perfected—from the precise alignment of carbon fiber panels to the subtle stitching patterns on the leather interior. Buyers were able to work directly with Aston Martin to tailor every aspect of the car, from paint shades that were mixed to unique specifications to interior leather and trim combinations that reflected their personal taste.

This bespoke approach made each One-77 **truly unique**, elevating it beyond a mere vehicle into a work of art. The hand-crafted nature of the car reinforced Aston Martin's reputation for excellence and ensured that the One-77 was more than just a hypercar—it was a personal expression of luxury, performance, and individuality.

Owning a One-77 is not simply about driving; it is about becoming part of a highly exclusive community. Buyers of the car are offered an intimate experience with the brand, including personalized consultations with engineers, designers, and technicians. Every interaction reinforces the car's sense of **prestige** and ensures that owners are intimately connected with the creation process.

The One-77's exclusivity also impacts its value as a collector's asset. With just seventy-seven units in existence, prices for well-maintained examples have soared since the car's introduction. For collectors, the One-77 is not only a **luxury possession** but also a tangible investment, a rare opportunity to own a piece of automotive history. The vehicle's scarcity, combined with its engineering sophistication and design brilliance, ensures that it holds a special place in the pantheon of hypercars.

Beyond its monetary value, the One-77 represents a lifestyle. It is a car meant to be cherished, driven selectively, and appreciated as a fusion of art, performance, and engineering mastery. Owners often describe a profound connection to the car, an emotional attachment that goes far beyond its technical capabilities.

While its exclusivity makes the One-77 desirable, its **performance** is what truly defines it. The naturally aspirated 7.3-liter V12 engine propels the car with an intensity that few vehicles can match. Acceleration is instantaneous yet controllable, delivering raw speed while maintaining driver confidence. The engine's torque curve is meticulously tuned to provide both explosive launches and smooth power delivery across the rev range, making every drive a sensory experience.

The One-77's suspension system and chassis geometry allow it to corner with astonishing precision. Double-wishbone suspension front and rear, adaptive dampers, and a rear-mounted transaxle provide a near-perfect balance between high-speed stability and nimble agility. Drivers describe a sensation of connectedness, as if the car is an extension of their own senses, responding instantly to every input. The combination of light weight, carbon fiber rigidity, and aerodynamic sophistication ensures that the One-77 remains planted at high speeds, while still offering comfort on less-than-perfect roads.

Even braking performance reinforces the car's track-capable dynamics. Carbon-ceramic brakes allow for repeated, aggressive stops without fade, offering unparalleled confidence during high-speed runs or spirited mountain drives. Every aspect of the One-77's performance package is engineered to provide an exhilarating yet controlled driving experience, bridging the gap between luxury grand tourer and track-ready hypercar.

Inside the One-77, Aston Martin's **commitment to luxury** is on full display. The cabin is a harmonious blend of opulent materials and advanced technology, offering an environment that is both comfortable and driver-focused. Every surface is meticulously finished, from the hand-stitched leather seats to the brushed aluminum controls and exotic wood trim. The ergonomics are carefully calibrated to ensure that controls fall naturally to hand, creating an intuitive, immersive driving environment.

Technology is seamlessly integrated. The bespoke infotainment system allows for full control of climate, navigation, and audio without compromising the tactile elegance of the cabin. Advanced driver assistance systems are unobtrusive yet effective, ensuring that performance is never compromised by modern conveniences. Even elements such as seating ergonomics, headrest support, and interior acoustics are engineered for an optimal balance between comfort and performance immersion.

Customization options were virtually limitless. Owners could select every detail—from leather stitching patterns and color contrasts to unique paint hues and bespoke interior veneers. This level of **personalization** ensured that each One-77 reflected the owner's tastes, making the car not only a performance marvel but also a statement of individuality.

The One-77 was more than a car; it was a benchmark. By combining **engineering mastery**, groundbreaking design, and uncompromising luxury, Aston Martin created a vehicle that would influence future hypercars for years to come. Lessons learned from the One-77's development informed subsequent models, including the Aston Martin Vulcan, Valkyrie, and the modern DBS and DB11 series. The hypercar served as a laboratory for new materials, powertrain technologies, and aerodynamic innovations that continue to resonate across the brand.

Its influence extends beyond engineering. The One-77 reinforced Aston Martin's design language and craftsmanship standards. Its flowing lines, aggressive stance, and meticulous interior detailing set a precedent for what customers would come to expect from future flagship models. In essence, the One-77 demonstrated that performance, exclusivity, and luxury could coexist at the highest levels—a philosophy that continues to shape Aston Martin's approach to modern hypercars.

The One-77 also had a profound impact on Aston Martin's position in the global automotive market. Its release reaffirmed the brand's reputation for **world-class engineering**, craftsmanship, and design excellence. It positioned Aston Martin alongside the very best hypercar manufacturers, such as Ferrari, Pagani, and McLaren, proving that the brand could innovate at the highest levels of automotive technology.

Beyond immediate performance accolades, the One-77 elevated Aston Martin's prestige among collectors and enthusiasts. Its limited production, bespoke craftsmanship, and unparalleled driving dynamics created a halo effect, enhancing the desirability of the brand's entire lineup. The car became both a benchmark and a symbol—a tangible representation of Aston Martin's philosophy: to create cars that are as beautiful to behold as they are exhilarating to drive.

The Aston Martin One-77 remains a **milestone in hypercar history**, representing the pinnacle of engineering, design, and exclusivity. Its naturally aspirated V12 engine, carbon fiber monocoque chassis, adaptive aerodynamics, and bespoke interior make it a triumph of modern automotive craftsmanship. Every element reflects Aston Martin's dedication to creating a car that excels not only in performance but also in luxury and individuality.

While only a select few will ever own one, the One-77's influence extends far beyond its seventy-seven units. It has inspired future hypercars, informed design philosophies, and reinforced Aston Martin's commitment to innovation and excellence. More than just a car, the One-77 is a **symbol of aspiration**, a fusion of art and engineering, and a testament to what is possible when a manufacturer dares to dream without compromise. Its legacy ensures that the One-77 will remain an icon of automotive history for decades to come.

11: THE MODERN VANTAGE

The Aston Martin Vantage has long held a revered place in the pantheon of sports cars, celebrated for its perfect harmony of **raw power, refined craftsmanship, and elegance**. From its earliest incarnation in 1951 to the latest 2018 iteration, the Vantage has epitomized the brand's unwavering commitment to delivering a driving experience that is as emotionally thrilling as it is technically sophisticated. While many manufacturers have chased performance at the expense of refinement, Aston Martin has built a legacy where engineering excellence and luxurious comfort coexist seamlessly. The Vantage nameplate represents more than a car; it embodies decades of innovation, precision engineering, and a unique aesthetic that sets Aston Martin apart from any competitor.

Unlike many sports cars that age poorly over decades, the Vantage has evolved intelligently, adapting to changing automotive demands while preserving its essential identity. Every generation reflects the zeitgeist of its era—blending the latest technologies with the timeless craftsmanship Aston Martin is renowned for. Yet, despite the passing decades, one constant remains: the Vantage is built to deliver an intimate connection between machine and driver, an exhilarating experience that rewards both skill and instinct.

The original Vantage was introduced in 1951 as a high-performance variant of the DB2, marking the first use of the "Vantage" designation—a term derived from motor racing, denoting a competitive edge. Aston Martin's engineers sought to create a car that was faster, more responsive, and more powerful than the standard DB2, without sacrificing elegance. The DB2 Vantage featured enhancements to its engine, improved suspension geometry, and subtle aerodynamic refinements, allowing drivers to experience a sports car that could command both the track and the road.

Performance, however, was only part of the story. Aston Martin's design ethos ensured that even the earliest Vantage was a statement of sophistication. The car's long, sweeping bonnet, elegant proportions, and attention to detail made it immediately recognizable as a British sports car of distinction. This balance between athleticism and refinement became the hallmark of the Vantage name, and it established the blueprint for future generations of the model. Early Vantage owners found themselves at the intersection of luxury and competition, enjoying a car that could accelerate heartbeats and turn heads simultaneously.

The 1970s brought a new era for Aston Martin and the Vantage in particular. The introduction of the 1977 V8 Vantage, often dubbed the "British Muscle Car," marked a significant shift. Powered by a 5.3-liter V8 engine producing over 280 horsepower, this iteration offered blistering performance in a package that was unmistakably aggressive. Its flared fenders, wide wheel arches, and bold stance conveyed a sense of muscularity previously unseen in Aston Martin models. Yet, unlike the purely performance-focused muscle cars of America, the V8 Vantage maintained the luxurious interior appointments that distinguished British craftsmanship.

By the 1980s and 1990s, Aston Martin had refined the Vantage further, integrating modern technologies while retaining the classic feel of the car. The 1993 Virage V8 Vantage, for example, brought more contemporary engineering to the fore, including better suspension geometry, improved braking systems, and a more sophisticated cabin layout. Although it lacked the raw fame of the earlier V8, the Virage V8 laid the groundwork for Aston Martin's future by showing that performance could coexist with comfort and advanced electronics. The Vantage had transitioned from being a racing-inspired upgrade to a fully-fledged luxury sports car, ready to compete with a new generation of international competitors.

The early 2000s marked a renaissance for Aston Martin, with new ownership, renewed investment, and a strategic focus on broadening the brand's appeal. It was in this era that the 2005 V8 Vantage emerged, representing a radical departure from its predecessors. Smaller, lighter, and more dynamically focused than previous models, it redefined the Vantage as a contemporary sports car capable of thrilling performance while remaining usable in everyday life.

Under its sculpted bonnet lay a 4.3-liter V8 engine, producing 380 horsepower and enabling the car to sprint from 0 to 60 mph in just 4.9 seconds. With its finely tuned suspension, precise steering, and light chassis, the V8 Vantage could compete with the Porsche 911 and other performance benchmarks while maintaining Aston Martin's distinctive

British elegance. Reviewers hailed it as the perfect synthesis of agility, power, and luxury—a driver's car in the truest sense.

The interior was meticulously designed, featuring hand-stitched leather, polished metal accents, and ergonomically optimized controls. Unlike earlier sports cars that compromised comfort for performance, the 2005 V8 Vantage offered an interior that invited long journeys and daily use without diminishing its sports credentials. In many ways, this Vantage reestablished the brand's credibility in the modern sports car segment and attracted a new generation of buyers eager for performance married to refinement.

One of the most critical innovations in the 2005 V8 Vantage was its aluminum-intensive chassis. By integrating lightweight materials without sacrificing strength, Aston Martin engineers created a car that felt agile yet composed, balancing responsiveness with stability at speed. This approach enabled precise handling and ensured that power delivery from the V8 engine was maximized through reduced inertia. The V8 Vantage's braking system, featuring carbon-ceramic discs, further underscored Aston Martin's commitment to high-performance engineering capable of meeting modern demands.

Suspension technology also received a significant overhaul. A double-wishbone setup at both the front and rear allowed for optimized handling dynamics, maintaining tire contact and grip under cornering forces that would challenge lesser chassis designs. These engineering advancements ensured that the V8 Vantage was not just a straight-line performer but also a car capable of delivering confidence-inspiring agility on twisty roads or racetrack circuits.

In 2009, Aston Martin expanded the Vantage range with the introduction of the V12 Vantage. Featuring a naturally aspirated 5.9-liter V12 engine producing 510 horsepower, the V12 variant offered a performance leap that transformed the driving experience. Acceleration from 0 to 60 mph dropped to just 4.2 seconds, and the car could exceed 190 mph, giving drivers a supercar-level experience in a relatively compact chassis.

Beyond raw power, the V12 Vantage introduced several enhancements: larger brakes, an upgraded suspension for more aggressive handling, and improved aerodynamic elements to increase downforce at speed. The exhaust system, finely tuned to complement the V12's character, delivered a rich, throaty note that became instantly identifiable among enthusiasts. In this form, the Vantage became a car that could rival some of the most potent grand tourers while maintaining Aston Martin's signature blend of luxury and refinement.

Aston Martin ushered in a bold new era with the 2018 Vantage, a model that redefined the brand's approach to sports cars for the modern age. Built on a new platform shared with the DB11 yet designed with a more aggressive, athletic character, the 2018 Vantage embodies a perfect balance of **heritage, power, style, and technology**. At its heart lies a 4.0-liter twin-turbocharged V8 engine, developed in collaboration with Mercedes-AMG, producing a staggering 503 horsepower and 505 lb-ft of torque. This power allows the car to accelerate from 0 to 60 mph in just 3.5 seconds, putting it squarely in the hyper-performance bracket while retaining Aston Martin's signature refinement.

The new Vantage also features an eight-speed automatic transmission, replacing the previous seven-speed manual, with quicker, smoother gear shifts and optimized efficiency. Drivers still have access to a manual-like experience through paddle shifters, giving the car the perfect combination of speed, control, and driver engagement. The adaptive suspension system, recalibrated for both comfort and track performance, ensures that the Vantage remains agile on tight bends yet composed on highways—a testament to Aston Martin's relentless engineering discipline.

The 2018 Vantage's design embodies Aston Martin's philosophy of blending **athleticism, elegance, aerodynamic efficiency, and visual drama**. While drawing inspiration from the DB11 and classic Vantage lines, the car introduces sharper angles, a more muscular stance, and an aggressive front grille that demands attention. Wide wheel arches, sculpted fenders, and a low roofline not only enhance aesthetics but also optimize airflow and downforce. Every curve, vent, and crease is intentional, contributing both to visual impact and aerodynamic performance.

The rear of the car reinforces this aggressive posture, featuring a pronounced diffuser, quad exhaust outlets, and an integrated spoiler that enhances high-speed stability. Inside, the Vantage cabin combines hand-stitched leather, Alcantara, and brushed aluminum with cutting-edge technology, creating a cockpit that is simultaneously luxurious, functional, and driver-centric. Ergonomics are carefully considered: controls are positioned for intuitive access, while the infotainment system provides connectivity and navigation without distraction. The modern Vantage proves that high performance need not compromise refinement.

The collaboration with Mercedes-AMG has been instrumental in refining the 2018 Vantage's engine and electronics. AMG's expertise in turbocharged V8s ensures that power delivery is instantaneous, linear, and responsive, while maintaining Aston Martin's characteristic exhaust sound—a deep, resonant growl that excites the senses. This collaboration also facilitated advanced electronic aids, including dynamic stability control, torque vectoring, and selectable driving modes, which allow the Vantage to adapt seamlessly to varying conditions, from city streets to the racetrack.

The car's chassis benefits from meticulous tuning to ensure the twin-turbo V8's performance is harnessed effectively. Weight distribution, suspension geometry, and electronic damping systems are optimized to provide razor-sharp handling and exceptional cornering confidence. Unlike many modern performance cars, the 2018 Vantage preserves an analog connection between driver and road; feedback through the steering wheel and pedals remains tangible and engaging, preserving the essence of a true sports car experience.

In addition to raw performance, Aston Martin equipped the modern Vantage with a suite of technologies designed to enhance **safety, comfort, connectivity, and driving satisfaction**. Advanced driver assistance systems, such as adaptive cruise control and collision mitigation, are integrated unobtrusively, maintaining a driver-focused environment while providing peace of mind. The infotainment system, featuring a high-definition touchscreen, navigation, and premium audio, ensures that long journeys are as comfortable and connected as they are thrilling.

The integration of adaptive lighting, aerodynamic enhancements, and traction management systems highlights Aston Martin's commitment to blending traditional craftsmanship with modern performance engineering. Each feature, from electronically controlled rear diffusers to high-performance brakes, is designed to complement the car's overall driving character, delivering an experience that is both intuitive and exhilarating.

The modern Vantage occupies a strategic position in Aston Martin's lineup, bridging the gap between the flagship DBS Superleggera and the more accessible, yet equally refined, DB11. It offers a more focused, driver-oriented experience, appealing to enthusiasts who prioritize agility and responsiveness without sacrificing comfort or luxury. Its combination of manageable dimensions, lightweight construction, and high-output V8 power makes it versatile for everyday driving as well as spirited track use.

Furthermore, the Vantage has helped Aston Martin expand its appeal to a younger, performance-minded audience. Its contemporary design, accessible yet exhilarating performance, and modern technological features make it a compelling alternative to traditional supercars and high-end sports cars. By occupying this niche, the Vantage reinforces Aston Martin's image as a brand that balances **heritage, innovation, exclusivity, and driving passion**.

Aston Martin has signaled its commitment to innovation and sustainability, indicating that future Vantage models may incorporate hybrid or fully electric powertrains. The challenge will be to preserve the car's distinctive driving character while integrating electrification. The brand's engineers are already exploring lightweight battery technologies, electric motor integration, and energy recovery systems to ensure that future models maintain the same dynamic prowess, refined comfort, and exclusivity that define the Vantage name.

This evolution reflects a broader trend in the automotive industry, where performance, efficiency, and sustainability must coexist. For Aston Martin, the goal is clear: maintain the heart and soul of the Vantage while embracing modern technologies that ensure the model remains relevant for the next generation of drivers. The fusion of electric innovation with traditional Aston Martin craftsmanship promises an exciting future, one where the Vantage continues to define the brand's sports car identity.

Through seven decades of evolution, the Vantage has consistently exemplified **performance, elegance, innovation, engineering excellence, and exclusivity**. From the DB2 Vantage to the 2018 modern iteration, it has remained a benchmark for sports cars, combining heart-pounding performance with exquisite craftsmanship. Its design has evolved without losing its quintessential Aston Martin DNA: long bonnets, aggressive stance, flowing lines, and interiors that envelop drivers in luxury and precision.

The Vantage's enduring popularity demonstrates that Aston Martin can evolve with the times while staying true to its core values. It remains a symbol of British engineering prowess, a car that excites the senses, challenges expectations, and rewards skillful driving. As it continues to develop, the Vantage promises to remain at the forefront of Aston Martin's sports car lineup, offering a compelling blend of heritage, innovation, and unmatched driving pleasure.

The Aston Martin Vantage is far more than a sports car; it is a statement of intent. It encapsulates decades of technological advancement, design evolution, and performance engineering excellence, while offering an emotional and visceral connection between car and driver. The 2018 model, with its AMG-enhanced V8, adaptive chassis, advanced electronics, and luxurious cockpit, represents the culmination of years of refinement, while also pointing toward a future of sustainable innovation.

Ultimately, the Vantage embodies Aston Martin's philosophy: the pursuit of excellence, the blending of power and elegance, and the creation of cars that are as thrilling to drive as they are beautiful to behold. Whether on the open road, a winding track, or in the hands of collectors, the Vantage remains a timeless icon—proof that performance and luxury can exist in perfect harmony, and that Aston Martin continues to define the pinnacle of British automotive artistry.

12: INNOVATION AND TECHNOLOGY

Aston Martin has long been celebrated as a brand that seamlessly intertwines luxury with performance, yet its distinguishing hallmark has always been technological innovation. From the earliest DB models to today's hybrid and electric prototypes, the company has consistently pushed the boundaries of what is possible in automotive engineering. Its philosophy is straightforward yet ambitious: to craft cars that not only dazzle with aesthetics and speed but also embody the cutting edge of mechanical and electronic innovation. In a world where technological advancement evolves at breakneck speed, Aston Martin has maintained its identity as a purveyor of elegant, high-performance machines that merge heritage with modernity. This chapter explores the intricate tapestry of Aston Martin's technological evolution, from pioneering early developments to future-ready innovations that promise to reshape the landscape of luxury performance automobiles.

Aston Martin's story of technological progress begins in the formative years of the brand. While competitors were often content with conventional designs and mechanical configurations, Aston Martin recognized the advantage of embracing emerging technologies. The DB series, particularly the DB2 introduced in 1949, exemplifies this forward-thinking approach. It was not merely a car—it was a statement of intent. Lightweight aluminum panels, an advanced inline-six engine, and refined suspension marked a turning point in the luxury sports car market. The DB2 illustrated that a car could simultaneously deliver exceptional handling, aerodynamic efficiency, and fuel economy without compromising the elegance that defined Aston Martin. These early innovations laid the groundwork for decades of engineering excellence, establishing a precedent for integrating technology with bespoke craftsmanship.

By focusing on materials science and aerodynamics, Aston Martin engineers ensured that each model would push performance boundaries while maintaining a refined driving experience. Aluminum bodies replaced heavier steel, reducing overall mass and enhancing agility, while precision-engineered suspensions offered greater stability without sacrificing comfort. These incremental innovations set Aston Martin apart in a rapidly evolving automotive landscape, allowing the brand to appeal to a niche clientele that valued both performance and sophistication. The marriage of lightweight materials and performance-focused engineering became a signature Aston Martin trait, one that remains central to the brand's philosophy today.

One of the most transformative technological advancements in Aston Martin's history is the adaptive suspension system. Introduced in models such as the DB9 and later refined in the V8 Vantage and DB11, adaptive suspension revolutionized the driving experience by allowing the car to respond in real-time to road conditions and driver inputs. Gone were the days when handling and comfort were mutually exclusive; this system allowed for both. Through a network of high-precision sensors, the vehicle monitors variables including road surface irregularities, speed, and lateral forces. This data feeds into the onboard electronic control unit, which continuously adjusts damping forces and suspension stiffness. The result is a ride that is at once composed and agile—luxurious on highways, yet responsive on curving mountain roads.

The benefits of adaptive suspension extend beyond comfort. By continuously optimizing wheel-to-road contact, the system enhances traction and stability, crucial in a high-powered, rear-wheel-drive sports car. Engineers have meticulously calibrated the software to balance performance with refinement, ensuring that adjustments occur seamlessly without disrupting the driver's experience. In vehicles like the V8 Vantage, the adaptive suspension delivers precise cornering with minimal body roll while maintaining plushness over uneven surfaces. Aston Martin's implementation of this technology represents a holistic approach to vehicle dynamics, where mechanical and electronic systems are harmoniously integrated to elevate both driver confidence and enjoyment.

With the rise of environmental awareness and stringent emissions regulations, Aston Martin faced the challenge of marrying sustainability with performance. The solution lay in hybridization—a technology that allows for cleaner, more efficient propulsion without compromising the visceral driving experience. The RapidE concept car, unveiled in 2015, was Aston Martin's first tangible step into the realm of electric mobility. Though never produced en masse, the RapidE demonstrated that the brand could integrate electric powertrains without diluting the character and luxury for which it is renowned. This experiment set the stage for a broader commitment to electrification, blending internal combustion engines with electric motors to create hybrid powertrains that offer both performance and environmental responsibility.

The forthcoming DBX hybrid illustrates Aston Martin's vision for the future. Combining a traditional V8 or V12 engine with electric assist, the hybrid system improves fuel efficiency and reduces emissions while enhancing acceleration and torque delivery. Unlike conventional hybrids, Aston Martin's approach focuses on driving engagement, ensuring that the electric components complement rather than overshadow the signature engine character. By leveraging hybrid technology, the brand anticipates stricter regulatory landscapes while continuing to satisfy customers' appetite for high-performance, luxury vehicles. In this way, Aston Martin demonstrates that innovation does not necessitate compromise—it is an opportunity to enhance both efficiency and exhilaration.

Aston Martin's electrification strategy goes beyond hybrids. By 2030, the brand has committed to transitioning its entire lineup to electric or electrified vehicles, signaling a fundamental shift in automotive design philosophy. Electric motors offer instant torque delivery, a characteristic that complements Aston Martin's performance ethos. The challenge, however, lies in preserving the auditory and tactile experience associated with internal combustion engines—the roar of a V12, the precise engagement of a manual gearbox. Engineers have approached this challenge by combining battery-powered propulsion with advanced software and sound engineering, ensuring that electric models maintain the brand's emotional appeal.

Moreover, electric vehicles enable Aston Martin to experiment with innovative chassis layouts and aerodynamic designs. The absence of a traditional engine allows for lower center-of-gravity placement, improved weight distribution, and more flexible cabin packaging. These developments promise not only environmental benefits but also enhanced driving dynamics, aligning with Aston Martin's tradition of performance excellence. In this sense, electrification is not a departure from heritage but an evolution—a way to redefine luxury performance in a sustainable world.

Luxury today extends beyond raw performance; it encompasses connectivity, convenience, and digital sophistication. Recognizing this, Aston Martin has invested heavily in infotainment technology. Early models featured basic audio systems and rudimentary navigation, but modern vehicles such as the DB11 incorporate fully integrated, high-resolution touchscreens, voice command functionality, and seamless smartphone connectivity. These systems allow drivers to manage navigation, climate control, media, and vehicle settings intuitively, minimizing distraction and enhancing the overall driving experience. Aston Martin's approach is to integrate technology without disrupting the elegance of the cabin, ensuring that each interface complements the tactile luxury of leather, wood, and aluminum.

In addition, advanced driver-assistance systems (ADAS) are now an integral part of Aston Martin vehicles.

Lane-keeping assist, adaptive cruise control, and emergency braking enhance safety without diluting the driving experience. The brand carefully calibrates these systems to be assistive rather than intrusive, maintaining the engagement that enthusiasts expect from a sports car. The result is a vehicle that is simultaneously sophisticated, safe, and exhilarating—a fusion of tradition and modernity.

As the automotive industry moves toward autonomy, Aston Martin has explored semi-autonomous technologies with a philosophy rooted in driver engagement. Unlike fully autonomous vehicles, which can remove the human element entirely, Aston Martin focuses on augmenting the driver rather than replacing them. Semi-autonomous features, such as highway autopilot, adaptive cruise control, and intelligent lane guidance, are integrated to provide convenience during long journeys without compromising the thrill of driving. These systems employ sophisticated sensors, radar, and cameras to monitor surrounding vehicles, road conditions, and driver behavior, ensuring both safety and performance.

Aston Martin engineers emphasize subtlety and precision in these technologies. For instance, the adaptive systems automatically adjust steering inputs and throttle response to maintain optimal handling, yet remain transparent to the driver unless intervention is required. The goal is to provide a safety net rather than dictate the driving experience. By maintaining the driver at the center of the equation, Aston Martin ensures that autonomous features enhance confidence while preserving the visceral connection to the car—a hallmark of the brand's identity.

The transition to full electrification represents one of the most ambitious phases in Aston Martin's history. Unlike conventional internal combustion engines, electric powertrains deliver instantaneous torque, which radically changes acceleration dynamics and responsiveness. Aston Martin has harnessed this advantage while ensuring its cars retain the brand's signature character. Electric motors are paired with lightweight chassis engineering and adaptive suspension, producing vehicles that are faster, more agile, and environmentally responsible. Battery placement is strategically engineered to lower the center of gravity, enhancing handling while maintaining luxurious cabin space.

In upcoming models, such as the fully electric Rapide-E successor, Aston Martin plans to blend high-performance electrification with bespoke craftsmanship. Each vehicle will integrate cutting-edge thermal management systems to ensure optimal battery performance while minimizing heat-related degradation. Advanced regenerative braking systems will enhance energy efficiency without compromising braking feel, a critical aspect of maintaining the tactile driving experience. Aston Martin views electrification not merely as compliance with environmental regulations but as a new frontier for engineering innovation—a way to redefine what it means to drive a luxury performance car in the 21st century.

Artificial intelligence is gradually becoming an integral component of Aston Martin's technological strategy. AI algorithms analyze real-time data from vehicle sensors to optimize performance and safety continuously. This includes predictive suspension adjustments, adaptive throttle calibration, and intelligent torque distribution to ensure that each driving scenario is handled with maximum precision. AI also enables advanced driver-assistance features to learn from user behavior, personalizing the driving experience while anticipating road hazards.

For example, in future Vantage and DBX models, AI will actively monitor driving habits, adjusting vehicle settings proactively. If a driver favors aggressive cornering, the system can subtly adjust traction control and damping to enhance responsiveness. Conversely, during relaxed highway cruising, the car automatically prioritizes comfort and efficiency. This dynamic adaptability exemplifies Aston Martin's ability to merge cutting-edge technology with its traditional focus on driver enjoyment.

Aston Martin is also pioneering vehicle connectivity, allowing cars to interact with digital infrastructure and other vehicles seamlessly. These connected systems provide real-time traffic updates, predictive navigation, and enhanced safety alerts. By leveraging cloud-based software, vehicles can receive over-the-air updates for performance optimizations, infotainment enhancements, and security improvements. This continuous evolution ensures that Aston Martin cars remain at the technological forefront, without requiring owners to trade in their vehicles for the latest innovations.

Inside the cabin, connected technology enhances convenience and luxury. From voice-activated climate control to AI-driven infotainment recommendations, the cabin becomes a sophisticated ecosystem tailored to the driver's preferences. Aston Martin integrates these systems unobtrusively, ensuring that luxury materials—leather, wood, and aluminum—remain the tactile focus, while digital interfaces provide seamless control and intuitive operation. This delicate balance between tradition and modernity is emblematic of Aston Martin's philosophy: technology serves the experience, never dominates it.

Technological innovation at Aston Martin extends beyond propulsion and electronics to the materials used in construction. The company has increasingly incorporated sustainable and lightweight materials into its vehicles, such as recycled carbon fiber, responsibly sourced leather, and bio-based composites. These materials not only reduce environmental impact but also enhance performance through weight reduction and improved structural rigidity.

In models like the DB11 and future Vantage variants, interior panels are crafted using eco-conscious alternatives without sacrificing aesthetic appeal or tactile quality. Aston Martin's engineers work alongside designers to ensure that sustainability complements luxury rather than undermines it. Even the paint processes have evolved to minimize volatile organic compounds, exemplifying the brand's holistic approach to responsible manufacturing. Sustainability and innovation, in this context, are intertwined, reinforcing Aston Martin's commitment to excellence while responding to the demands of a greener automotive future.

Looking beyond electrification and AI, Aston Martin's future hypercars are poised to combine traditional high-performance engineering with state-of-the-art technologies. The next generation of flagship models will feature hybrid-electric drivetrains, advanced aerodynamics with active control surfaces, and AI-assisted chassis management for unrivaled precision. These vehicles will not only exceed the performance of current models but also set new benchmarks in driver engagement, handling, and environmental responsibility.

Active aerodynamics, a feature pioneered in the One-77 and DBS Superleggera, will be further refined with adaptive computational systems. Spoilers, diffusers, and winglets will adjust dynamically to optimize downforce, reduce drag, and enhance cornering grip. Combined with lightweight construction and high-output electrified engines, future Aston Martin hypercars promise a driving experience that is faster, more precise, and more exhilarating than ever before—all while maintaining the timeless elegance expected from the marque.

Despite these technological advancements, Aston Martin remains steadfast in its commitment to heritage. Every innovation is carefully integrated to preserve the essence of what makes an Aston Martin uniquely luxurious and exhilarating. Whether it is the tactile feedback of a steering wheel, the symphonic roar of a V8 or V12 engine, or the meticulous hand-stitched leather interior, the brand ensures that technology serves the driver's experience rather than replacing it.

This balance between innovation and tradition is evident in vehicles like the DB11, Vantage, and future electric models. Advanced electronics, hybrid powertrains, AI systems, and sustainable materials are all incorporated in a manner that enhances rather than diminishes the visceral, human-centered experience. Aston Martin's engineers view technology as a tool to amplify performance, comfort, and exclusivity, ensuring that every car retains the emotional connection that has defined the brand for over a century.

Aston Martin's journey in innovation is a testament to the brand's foresight, adaptability, and unwavering pursuit of excellence. From the early aluminum-bodied DB series to hybrid-powered DBX SUVs and electric hypercars, Aston Martin has consistently integrated cutting-edge technologies while preserving the artistry and craftsmanship that define its identity. Adaptive suspension, hybrid powertrains, AI-assisted driving, connected infotainment, and sustainable materials illustrate a holistic approach to innovation—one that combines performance, luxury, and responsibility.

Looking forward, Aston Martin's commitment to electrification, autonomous systems, and next-generation hypercars ensures that the brand will remain at the forefront of luxury automotive technology. Each innovation reinforces the core philosophy: that true luxury is not merely about opulence, nor is performance solely about speed. Aston Martin demonstrates that technology, when carefully and thoughtfully integrated, can elevate the driving experience to unprecedented heights. In doing so, it secures its position as one of the world's most respected and visionary automotive brands—a company that honors tradition while shaping the future.

13: THE 'AM' CUSTOMIZATION AND PERSONALIZATION

Aston Martin has always been synonymous with **luxury, performance, craftsmanship, elegance, and exclusivity**, but the true jewel in the brand's crown is its unparalleled approach to personalization. Unlike most automakers, which offer standard packages with limited choices, Aston Martin gives its customers the ability to design a vehicle that is **entirely their own creation**, a car that is a reflection of their taste, identity, and lifestyle. This dedication to individuality is woven into the very DNA of the brand, making every Aston Martin more than just a high-performance automobile—it is a canvas for artistry, imagination, and personal expression.

From the moment a customer steps into the world of Aston Martin, they are presented with a unique opportunity: to take part in a process that transforms a vision into a rolling masterpiece. This philosophy is exemplified in the Aston Martin Personalization Program and, most notably, through the brand's exclusive Q by Aston Martin division. Here, the boundaries of bespoke automotive design are continually pushed, allowing clients to transcend conventional choices and create vehicles that are singularly their own. Every stitch, every curve, every hue is subject to the whims of the owner, ensuring that no two cars are alike.

The Personalization Program is more than an optional add-on; it is a philosophy, a meticulous process that allows owners to explore every element of their vehicle, from performance specifications to aesthetic details. Aston Martin understands that luxury is not simply about speed or prestige—it is about creating a deeply personal connection between a car and its driver. A bespoke Aston Martin is an extension of the owner's personality, their tastes, and their vision of perfection on four wheels.

For many luxury car manufacturers, customization is limited to preselected paint options, a handful of leather choices, and perhaps a choice of alloy wheels. Aston Martin approaches personalization differently. Every car begins as a blank canvas, allowing owners to participate in a **creative journey that shapes their dream vehicle** from the ground up. This philosophy ensures that the finished car is not merely a commodity—it is an individual statement of style, sophistication, and automotive artistry.

The process begins with a detailed consultation. Trained Aston Martin consultants guide each customer through the seemingly endless possibilities, from exterior colors and finishes to interior materials and stitching. Owners are encouraged to imagine their perfect car without constraints, with the reassurance that Aston Martin's expert team will make even the most ambitious visions a reality. Whether it's a unique color gradient for the exterior, a rare exotic leather for the cabin, or a personalized monogram stitched into the seats, every choice is considered carefully and executed with exacting precision.

Personalization is not simply about aesthetics. Aston Martin ensures that each option chosen enhances the driving experience, maintaining the brand's reputation for **exhilarating performance, impeccable comfort, and luxurious craftsmanship**. A bespoke vehicle does not compromise the engineering excellence for which Aston Martin is renowned. Instead, it elevates it, merging individuality with performance in a way that few manufacturers can match.

The exterior of an Aston Martin is often the first thing people notice, and it is here that the brand's personalization philosophy begins to shine. The Personalization Program offers an extraordinary palette of colors, finishes, and textures, ranging from classic metallics to daring, one-off hues developed specifically for the customer. Whether a deep emerald green, a silvery frost, or a bold, iridescent shade that shifts in the light, the exterior becomes a reflection of the owner's personality.

Aston Martin's commitment to perfection extends beyond color selection. Multiple layers of hand-applied paint are used to achieve depth, brilliance, and a flawless finish, resulting in a car that gleams under sunlight as if sculpted from liquid metal. Satin, matte, and pearlescent treatments provide additional options for subtlety or drama, allowing the vehicle to convey understated elegance or commanding presence. Even the smallest touches, such as custom trim pieces, badges, or carbon fiber accents, are tailored to the owner's vision, ensuring that every element is harmonious and intentional.

Some clients opt for fully bespoke designs, with exclusive graphics, stripes, or motifs integrated into the bodywork. These personalized elements are not merely decorative—they are a statement, a signature, and a testament to the collaboration between owner and artisan. In the hands of Aston Martin's master craftsmen, every car becomes a rolling work of art, celebrated as much for its beauty as for its speed and engineering.

Once the exterior is defined, the interior becomes the true playground of Aston Martin's Personalization Program. Here, craftsmanship, luxury, and creativity converge to create a cabin that is both refined and expressive. Every surface is customizable, from seats and dashboards to door panels, steering wheels, and even the pedals. The goal is to create a space that is a **perfect reflection of the owner's taste, style, and lifestyle.**

Materials are selected with painstaking care. Fine leathers sourced from across the globe, soft Alcantara, premium suede, exotic woods, polished metals, and carbon fiber—all are at the customer's disposal. The leather's color, texture, and grain can be matched or contrasted with stitching choices, embroidery, or piping, allowing for subtlety or bold statements. In some instances, Q by Aston Martin has produced interiors featuring rare materials such as meteorite inlays or precious metals, elevating the cabin into a space of unparalleled luxury.

Stitching is an often-overlooked but vital element of interior design. Aston Martin allows owners to choose thread colors and patterns, creating visual continuity or dynamic contrast across seats, dashboards, and trim. Custom monograms, logos, or symbols can be incorporated into headrests or panels, providing a unique identity for the car that resonates with the owner's personality. This attention to detail ensures that the interior is not only visually stunning but also emotionally engaging, offering a tactile and aesthetic experience that is both luxurious and intimate.

Ergonomics, comfort, and functionality are also part of the bespoke process. Seat shapes, padding, and support can be tailored to the owner's preferences, while layouts can accommodate special storage or technology requirements. Every interior decision is a blend of artistry and engineering, ensuring that bespoke does not come at the expense of performance or comfort.

While the Personalization Program allows owners to customize nearly every detail, Q by Aston Martin represents the **ultimate pinnacle of automotive exclusivity**. Q is a dedicated design studio where Aston Martin's most skilled artisans, engineers, and designers collaborate directly with clients to craft vehicles that go far beyond standard options. Here, imagination becomes reality, and no request is too bold.

Q by Aston Martin specializes in one-off vehicles, limited editions, and unique design elements that reflect the owner's tastes in every conceivable way. The division can create entirely new paint colors, interior designs, bespoke wheels, and even custom aerodynamic enhancements. Every car is a singular creation, treated as a masterpiece rather than a production model, where time, attention, and skill are devoted to achieving perfection in every detail.

The artisans at Q work with cutting-edge techniques alongside traditional craftsmanship. Hand-stitched leathers, hand-applied paints, custom metalwork, and precision-engineered components all come together to produce vehicles that are as functional as they are artful. These bespoke vehicles often incorporate features and innovations not available anywhere else in the world, reflecting the owner's personality and creating a level of exclusivity unmatched in the automotive industry.

Aston Martin's commitment to personalization extends beyond individual commissions into the creation of **limited-edition vehicles that exemplify exclusivity, rarity, and craftsmanship**. These models take the art of bespoke automotive design to a level that few other manufacturers can achieve. Produced in extremely small numbers, limited editions offer unique design elements, performance enhancements, and a sense of prestige reserved for the fortunate few. Each car tells a story, celebrating innovation while providing owners with a vehicle that is truly one-of-a-kind.

One of the most famous examples is the Aston Martin One-77, a hypercar that epitomizes both performance and artistic excellence. Produced in just seventy-seven units worldwide, the One-77 combines the power of a 7.3-liter V12 engine with the unparalleled beauty of bespoke craftsmanship. Every detail of the car, from its aerodynamically sculpted bodywork to the luxurious interior, was designed to reflect the pinnacle of Aston Martin's engineering and design capabilities. For collectors, the One-77 is more than a car—it is a statement, an investment, and a tangible representation of their personal taste.

Other limited-edition models have included special variants of the Vantage, DB11, and Vanquish. These vehicles often feature **exclusive paint finishes, custom interior materials, enhanced performance packages, and personalized badges**, highlighting their distinctiveness from standard production models. Some editions have even incorporated rare, one-off materials, such as exotic woods, carbon fiber composites, or precious metals, elevating the vehicle from a sports car to a rolling work of art. These limited editions not only serve as examples of Aston Martin's craftsmanship but also reinforce the brand's dedication to individuality, ensuring each car remains a singular expression of luxury.

Bespoke customization is not limited to aesthetics; it extends to **engine tuning, suspension calibration, and performance optimization**, allowing each vehicle to match the owner's driving style. Aston Martin engineers work closely with clients to develop vehicles that deliver not only breathtaking speed but also a driving experience tailored to the owner's desires. Whether it's a car optimized for the track, a comfortable grand tourer for long journeys, or a perfect balance of both, customization ensures that each vehicle performs as uniquely as it looks.

For instance, clients can choose adaptive suspension settings, brake specifications, and tire configurations that best suit their driving preferences. Even the software controlling traction, stability, and power delivery can be fine-tuned, providing a level of personalization rarely found in high-performance vehicles. Aston Martin ensures that bespoke performance does not compromise reliability, safety, or luxury, merging the thrill of driving with the refinement expected of the marque.

Aston Martin's approach to customization is more than a marketing strategy—it is a **philosophy rooted in the brand's identity and heritage**. By empowering clients to design their vehicles, Aston Martin transforms ownership into a deeply personal experience, creating a sense of emotional connection between the driver and machine. Every car becomes a reflection of the individual who commissioned it, whether through bespoke materials, personalized insignias, or one-of-a-kind design features. This philosophy ensures that owning an Aston Martin is as much about personal expression as it is about performance or luxury.

The legacy of Aston Martin personalization can be traced through its storied history. From early bespoke models in the 1950s, crafted to the specifications of discerning clients, to modern vehicles equipped with advanced technology and unique materials, the commitment to individuality has been unwavering. Each era of Aston Martin innovation reflects a dedication to combining engineering excellence with artistic freedom, ensuring that every vehicle is both a masterpiece and a personal statement.

Q by Aston Martin is the brand's ultimate expression of **bespoke craftsmanship, design innovation, creativity, and exclusivity**. While standard personalization allows for a wide range of options, Q pushes the boundaries further, creating vehicles that are truly unparalleled. Here, Aston Martin's artisans collaborate directly with clients to craft cars that may feature one-off body panels, exclusive paintwork, rare materials, and fully customized interiors. Every element is considered, from wheel design to dashboard layout, ensuring that the car is a singular expression of the owner's vision.

The Q division has undertaken projects that blend artistry with engineering precision, creating cars that are as unique as the clients who commission them. Some vehicles include hand-painted exteriors, rare leather combinations, intricate stitching, and custom badges that signify the car's individuality. In certain cases, Q has designed entirely new features—like specialized aerodynamic enhancements, exclusive trim options, or innovative cabin layouts—that go beyond conventional manufacturing capabilities. The result is a car that is not only functional and high-performing but also a living work of art, bearing the signature of its creator.

Within Q by Aston Martin, interiors become a canvas for **luxury, craftsmanship, creativity, personalization, and elegance**. Owners can choose every material, texture, and color for their cabin, crafting a space that mirrors their aesthetic sensibilities. From hand-stitched leather to Alcantara, carbon fiber to rare exotic woods, every surface is customizable. Stitching patterns, monograms, and subtle embossing allow for further personalization, ensuring that the cabin is as unique as the exterior.

Advanced technology is seamlessly integrated into these bespoke interiors. Infotainment systems, driver interfaces, and connectivity features are tailored to the owner's specifications, ensuring convenience and comfort without sacrificing aesthetic harmony. Even functional elements, such as seat ergonomics, pedal configurations, and storage solutions, can be customized, allowing each car to meet the precise demands of its owner. The result is an interior that balances luxury with practicality, elegance with individuality—a space where every journey becomes an experience.

Aston Martin's limited editions are not only a showcase of design ingenuity but also a testament to **collectibility, exclusivity, craftsmanship, rarity, and prestige**. These vehicles are produced in small quantities, making them highly sought after by collectors and enthusiasts. Limited editions often feature unique styling cues, one-off materials, and performance enhancements that distinguish them from standard models.

The One-77 remains the crown jewel of Aston Martin's limited editions, representing the epitome of engineering, design, and bespoke craftsmanship. Other notable examples include special editions of the Vantage, DB11, and Vanquish, each offering unique enhancements that cater to discerning clients. By producing these ultra-rare vehicles, Aston Martin preserves the exclusivity and desirability of its brand while demonstrating its ability to craft vehicles that are both exceptional in performance and artistic expression.

At the heart of Aston Martin's personalization philosophy is the idea that **every car should be a masterpiece, individually crafted for its owner**. The Personalization Program and Q division empower clients to create vehicles that reflect their vision of perfection. Each car is a combination of innovation, design, engineering, and craftsmanship, with no detail overlooked.

Whether through bespoke exterior finishes, hand-selected interior materials, personalized stitching, or one-off features, Aston Martin ensures that every vehicle is a true reflection of its owner's personality. The result is a car that is not only a high-performance machine but also a work of art—an expression of identity, style, and taste.

As automotive technology evolves, Aston Martin continues to expand the possibilities for **personalization, exclusivity, creativity, design, and innovation**. Emerging technologies, such as advanced materials, digital design tools, and additive manufacturing, allow for even greater flexibility in creating bespoke vehicles. Owners of future Aston Martins will be able to explore combinations of finishes, materials, and performance options previously unimaginable.

Personalization will also extend into the digital realm, with integrated infotainment, connected services, and intelligent vehicle systems allowing for unique experiences tailored to the owner. Aston Martin's commitment to creating one-of-a-kind vehicles will endure, ensuring that every new model can be a canvas for expression, combining **heritage, innovation, luxury, and performance** in perfect harmony.

Aston Martin's dedication to personalization has cemented its place as a leader in luxury automotive craftsmanship. The Personalization Program and Q by Aston Martin division have redefined what it means to own a bespoke vehicle, allowing clients to create cars that are as individual as they are extraordinary. Every vehicle, from a limited-edition hypercar to a personalized Vantage, represents the perfect marriage of engineering excellence, artistry, and identity.

In a world increasingly dominated by mass production, Aston Martin stands apart, proving that true luxury lies in individuality. By offering unrivaled levels of customization and exclusivity, the brand ensures that each car is not just a machine—it is a story, a work of art, and a reflection of its owner. Aston Martin continues to shape the future of automotive personalization, ensuring that the legacy of bespoke luxury endures for generations to come.

Aston Martin's approach to customization and personalization is a testament to **luxury, individuality, artistry, innovation, and excellence**. Through its Personalization Program and Q by Aston Martin division, the brand empowers clients to create cars that transcend conventional expectations. From the paintwork to the interiors, from performance options to one-off features, every vehicle is crafted to perfection, ensuring that each Aston Martin is a true reflection of its owner.

As the brand moves into the future, the fusion of traditional craftsmanship with cutting-edge technology will continue to redefine the possibilities of bespoke automotive design. Aston Martin remains committed to creating cars that are not merely modes of transportation but personal statements, works of art, and embodiments of luxury, performance, and identity. Each vehicle is a testament to the brand's enduring philosophy: **crafting perfection, one car at a time.**

14: ASTON MARTIN'S HERITAGE AND CRAFTSMANSHIP

While Aston Martin's artistry is immediately visible in its hand-finished interiors and paintwork, the engineering underpinning each car is equally a form of craftsmanship. Every mechanical component, from the engine to the suspension, is designed with the same precision and attention to detail as the leather stitching and wood veneers. Engineers and artisans collaborate closely to ensure that performance, luxury, and aesthetics are fully integrated.

The engines, for example, are hand-assembled with meticulous attention to tolerances and balance. Each V12 or V8 engine is a symphony of components, engineered to deliver power, smoothness, and reliability. Pistons, crankshafts, and camshafts are inspected and polished by hand to ensure perfect fitment, and the assembly process is timed to guarantee precision. These mechanical masterpieces are not just engines—they are the beating heart of each Aston Martin, seamlessly blending technology and craftsmanship.

Inside an Aston Martin, every touch point demonstrates the philosophy of "luxury by hand." Artisans carefully apply fine leather to dashboards, steering wheels, and center consoles.

The grain of the leather is aligned with aesthetic intent, ensuring a harmonious visual and tactile experience. Even the choice of stitching pattern—double-stitch, cross-stitch, or bespoke monogramming—is considered down to the millimeter.

The cabin's wood trim is similarly treated as an artwork. Each veneer is selected for its grain, color, and tonal harmony, and is then hand-sanded, cut, and lacquered to create depth and warmth. Exotic materials such as carbon fiber, brushed aluminum, and rare metals are seamlessly incorporated, demonstrating Aston Martin's ability to blend traditional craftsmanship with cutting-edge technology. The result is a cabin where every detail feels intentional, luxurious, and timeless.

Aston Martin's philosophy extends beyond visible luxury; even the smallest, seemingly invisible components are crafted with care. Hinges, door handles, and fasteners are finished to exacting standards, ensuring that every interaction feels refined. Door closures are tuned for optimal feedback, creating a sense of solidity and precision. Every tactile experience, from shifting gears to adjusting a seat, is carefully considered to enhance the sense of quality.

Even the carpets and floor mats are hand-cut and stitched to match the interior's design theme. This attention to detail ensures that the Aston Martin experience is immersive, elegant, and cohesive. No component is overlooked, no matter how minor, because in the Aston Martin ethos, perfection is defined by the sum of its details.

The materials used in an Aston Martin are chosen not only for their visual appeal but also for their contribution to performance and longevity. Lightweight aluminum, carbon fiber, and high-strength alloys allow Aston Martin vehicles to remain agile, fast, and safe while retaining their luxurious presence.

Carbon fiber, for example, is incorporated in body panels, aerodynamic components, and interior trims. Its lightweight strength improves performance and fuel efficiency while providing a sculpted, futuristic aesthetic. Aluminum alloys are similarly used in chassis components to balance rigidity, crash safety, and responsive handling. By carefully selecting and hand-finishing these materials, Aston Martin ensures that every car combines functional engineering with visual beauty, creating a holistic driving experience.

Aston Martin's paint process is a world-renowned example of artisanal skill meeting modern precision. Each car may go through dozens of layers of paint, primer, and protective coating, with artisans carefully sanding, polishing, and inspecting each layer. The depth and luster of the final finish is not merely cosmetic—it communicates care, sophistication, and attention to detail.

Special finishes, such as satin, metallic, and pearlescent options, are applied to enhance light reflection and create visual depth. Custom, one-of-a-kind colors may be developed for individual clients, a testament to Aston Martin's commitment to personalization. This painstaking process ensures that every car emerging from the paint shop is a statement of elegance and craftsmanship.

The legacy of Aston Martin extends into the experience of commissioning a vehicle. Clients are invited to work with master artisans and designers to create a car that is fully reflective of their personal tastes. The process may involve selecting exterior colors, interior leathers, stitching styles, trims, wheel designs, and even custom engravings.

This bespoke approach is not merely decorative; it embodies Aston Martin's ethos that a car should be an extension of the owner's personality. The collaborative design experience ensures that each Aston Martin is as unique as the individual commissioning it, merging luxury, artistry, and exclusivity into one vehicle.

Aston Martin's dedication to craftsmanship is particularly evident in its limited-edition models. These vehicles often feature unique paint finishes, exclusive interiors, and enhanced performance components. Examples such as the One-77, Vanquish Zagato, and V12 Vantage S are celebrated for their combination of artistry and engineering excellence.

These limited editions are crafted with even greater attention to detail, often incorporating bespoke elements requested by individual owners. The result is a collection of vehicles that are not only rare and valuable but also represent the pinnacle of Aston Martin's hand-built philosophy. Each limited edition reinforces the brand's commitment to creating vehicles that are masterpieces both in performance and aesthetic appeal.

Owning an Aston Martin is not simply about transportation; it is an experience that immerses the owner in a world of heritage, craftsmanship, and personalized luxury. The brand's dedication to hand-built perfection and attention to detail transforms every journey into a statement of elegance and engineering excellence.

From the tactile satisfaction of opening a perfectly aligned door to the visual splendor of a hand-polished paint finish under sunlight, every moment is designed to evoke awe and appreciation. It is this consistent commitment to quality, artistry, and experience that has cemented Aston Martin's reputation as one of the world's most prestigious automotive brands.

Aston Martin's legacy is rooted in over a century of innovation, precision, and artisan skill. The company's philosophy marries modern engineering with hand-built artistry, creating vehicles that are both high-performing and timelessly beautiful. Each car embodies the brand's heritage, with every stitch, panel, and mechanical component reflecting decades of refinement and passion.

As the company moves into the future, this commitment to craftsmanship and heritage remains central. New technologies, materials, and performance capabilities will continue to be integrated, but always with the same dedication to artistry and detail. Aston Martin's cars will remain more than vehicles—they will be symbols of legacy, personal expression, and the art of hand-built automotive excellence.

At Aston Martin, craftsmanship is not an afterthought; it is the foundation upon which every vehicle is built. By combining traditional hand-finishing techniques with modern engineering innovation, the brand creates cars that are as timeless as they are dynamic. From the careful selection of materials to the meticulous attention to detail in every assembly step, Aston Martin continues to uphold a tradition of excellence that is unmatched in the automotive world.

Through this dedication to heritage, artistry, and precision, Aston Martin ensures that each new generation of cars remains a masterpiece—an embodiment of luxury, performance, and human craftsmanship that transcends time and trends. The Aston Martin experience is, and always will be, the perfect harmony of tradition, innovation, and art.

Aston Martin's craftsmanship extends far beyond aesthetics; it encompasses the precision engineering that ensures performance and reliability. Each engine, gearbox, and suspension system is designed with an artisan's care, combining cutting-edge technology with hand-finished perfection. For example, the V12 engines are not only powerful—they are meticulously balanced, polished, and calibrated to deliver both thrilling performance and refined smoothness.

The chassis is another area where craftsmanship shines. Aston Martin engineers hand-fit every suspension component, ensuring precise alignment for superior handling. Carbon-ceramic brakes, precision-machined from high-performance composites, are tested and refined to provide exacting stopping power. Even the exhaust systems are tuned for both acoustic excellence and optimal performance. Each mechanical element is a testament to the brand's philosophy: true luxury exists when engineering and artistry are inseparable.

Inside the cabin, craftsmanship reaches its zenith. Every leather surface is hand-cut, stitched, and inspected for flawless texture and alignment. Seats are sculpted to support the human body while remaining visually striking. Alcantara, suede, and exotic hides are meticulously selected for grain, color, and durability. Wood veneers are hand-sanded and polished to reveal natural patterns, while brushed aluminum and carbon fiber accents are integrated with geometric precision.

Stitching patterns are an art form in themselves. Contrasting thread, hand-applied in precise sequences, emphasizes curves and panels while creating subtle visual drama. Monograms, personalized embroidery, and custom trim enhance individuality. Even the smallest components—the steering wheel rim, the gearshift lever, or door handles—are crafted to satisfy tactile, visual, and ergonomic perfection. Aston Martin interiors are not just luxury; they are immersive experiences in refined artistry.

Aston Martin's hand-painted finishes are legendary. Multiple layers of primer, base coat, and clear lacquer are applied, with each stage carefully inspected and polished. This labor-intensive process creates a depth and richness unmatched by mass-produced vehicles. Special finishes—satin, matte, or metallic—add dimensionality and subtlety.

Custom colors, often developed exclusively for individual clients, demonstrate Aston Martin's commitment to uniqueness. Each bespoke paintwork is a collaboration between the artisan and the client, reflecting personal style while maintaining the brand's iconic aesthetic. Even under varying light, the surfaces appear alive, revealing gradients and reflections that highlight the car's sculpted curves.

The foundation of Aston Martin's craftsmanship lies in its materials. Premium leather, exotic wood, carbon fiber, and brushed metals are not just chosen for beauty—they enhance durability, performance, and tactile appeal. Full-grain leather is selected to ensure softness, suppleness, and resilience. Wood veneers are carefully graded for tone and pattern consistency, ensuring visual harmony throughout the cabin.

Aluminum and carbon fiber are chosen for lightweight strength, enabling precise handling and agility. Metals used for accents are polished, brushed, or treated to create contrast without compromising tactile quality. This combination of artistry and engineering ensures that every surface feels intentional, every material contributes to the whole, and the car is both functional and visually stunning.

Central to Aston Martin's legacy are the artisans themselves. Each employee—from leather stitchers to bodywork specialists—is a master in their field. Skills are often passed down through generations, preserving techniques that cannot be replicated by machines.

Leather workers shape and stitch seats, ensuring comfort and visual beauty. Woodworkers hand-select veneers, shaping each panel with meticulous care. Bodywork specialists align panels with sub-millimeter precision, smoothing curves to perfection. Even the assembly of engines is done with artisan oversight, combining mechanical knowledge with intuitive feel. This human touch is what transforms raw materials into automotive masterpieces.

Beyond standard luxury, Aston Martin offers clients the ability to craft a car that is truly their own. Through the Personalization Program and Q by Aston Martin, clients can influence every detail—from exterior color to interior stitching, wheel design, and even bespoke powertrain specifications.

These cars are not just vehicles—they are extensions of the owner's identity. Collaboration between designers, artisans, and clients ensures that no two cars are alike. Every decision is guided by taste, style, and function, resulting in an automobile that is both a high-performance machine and a personal statement of refinement.

Limited-edition models epitomize Aston Martin's philosophy of bespoke luxury. Cars such as the One-77, Vanquish Zagato, and V12 Vantage S are produced in restricted numbers, often featuring unique colors, hand-finished interiors, and bespoke design elements. These vehicles embody the brand's pinnacle of craftsmanship, blending artistry, engineering, and exclusivity.

Owners of these limited editions experience a sense of distinction rarely found in the automotive world. Each vehicle represents countless hours of labor, precision, and creative input—an enduring testament to Aston Martin's dedication to making every car extraordinary.

Aston Martin's mastery is evident in even the smallest details. The feel of a door handle, the feedback from pedals, and the way controls engage are all considered carefully. Fasteners, panels, and interior components are aligned with precise tolerances, ensuring that the car feels cohesive and perfectly engineered.

Even decorative touches serve a purpose, balancing aesthetics with ergonomics. This meticulous attention to detail reinforces the brand's philosophy: luxury is defined not only by what is visible but also by how seamlessly all components function together.

Aston Martin continuously balances its rich heritage with technological innovation. Modern materials, lightweight construction, and advanced performance engineering complement centuries-old craftsmanship traditions. While carbon fiber and adaptive suspensions represent the future, hand-stitched leather and meticulously polished veneers preserve the essence of the past.

This duality ensures that every new Aston Martin retains the soul of a handcrafted luxury car while benefiting from modern performance, safety, and convenience. The brand's artisans and engineers work in harmony to create vehicles that are simultaneously innovative, luxurious, and timeless.

Ultimately, owning an Aston Martin is about experiencing the intersection of heritage, craftsmanship, and performance. Every car embodies decades of refinement, artistry, and precision engineering. From the hand-stitched leather and polished veneers to the hand-assembled engines and meticulously applied paint, each element is a reflection of the brand's unwavering commitment to excellence.

The legacy of Aston Martin is not just in the vehicles themselves, but in the people who craft them—the artisans, designers, and engineers who combine skill, creativity, and passion to produce cars that are more than transportation: they are works of art.

Aston Martin's heritage and craftsmanship are inseparable from its identity. Each vehicle represents a culmination of art, engineering, and tradition. Hand-built interiors, meticulously engineered engines, and flawless finishes converge to create automobiles that are timeless, luxurious, and thrilling to drive.

In the ever-evolving automotive landscape, Aston Martin's dedication to craftsmanship ensures that every car remains a masterpiece—an enduring symbol of elegance, innovation, and the human touch. The legacy continues, as new generations of artisans, engineers, and designers uphold the brand's philosophy: to craft cars that are extraordinary in every conceivable way, and to make every Aston Martin a story of passion, precision, and perfection.

15: THE DBS SUPERLEGGERA

The DBS Superleggera is not just a machine—it is a demonstration of engineering excellence at Aston Martin's highest level. Beneath its sculpted bodywork lies a meticulously engineered chassis and drivetrain designed to extract the ultimate performance from the twin-turbo V12 engine. Every element, from the lightweight aluminum and carbon fiber construction to the precisely tuned suspension, has been conceived to ensure the car responds with immediacy and precision under every driving condition.

One of the most remarkable aspects of the DBS Superleggera's engineering is its aerodynamic sophistication. Active aerodynamics, including an adjustable rear spoiler and front splitter, adapt in real-time to optimize downforce and minimize drag. This ensures stability at high speeds while retaining the elegance of the car's profile. The airflow is carefully channeled to cool critical components, including the brakes and engine, ensuring consistent performance even under extreme conditions.

The car's suspension system exemplifies Aston Martin's commitment to marrying comfort with performance. With adaptive dampers, the DBS Superleggera seamlessly transitions between a relaxed grand touring ride and a razor-sharp, track-ready stance. The vehicle remains composed and planted, even when negotiating sharp bends, a testament to Aston Martin's painstaking development and testing protocols.

While its performance capabilities impress, the DBS Superleggera's interior reinforces Aston Martin's commitment to refined luxury. Every surface inside the cabin is the result of artisanal craftsmanship, using hand-stitched leather, polished carbon fiber, and brushed aluminum accents. This combination creates an environment where the tactile sensation of luxury matches the visual elegance of the exterior.

The seats are sculpted to provide maximum comfort during long-distance journeys while offering exceptional lateral support during spirited driving. Each stitch is hand-placed, reflecting Aston Martin's dedication to detail. Drivers and passengers alike experience a cabin that combines opulence with ergonomics, ensuring comfort is never sacrificed for performance.

Technology is integrated in a manner that enhances the driving experience without distracting from it. The infotainment system offers intuitive controls, navigation, and connectivity features, while the digital instrument cluster delivers essential performance data in a sleek, modern interface. Even the air vents and control surfaces are designed with precision, reinforcing the feeling that every detail has been deliberately considered.

A key factor in the DBS Superleggera's performance is its extensive use of carbon fiber and lightweight materials. Aston Martin has long embraced the principle of weight reduction as a route to enhanced agility and efficiency, and the DBS Superleggera exemplifies this philosophy. Components such as the roof, rear diffuser, and front splitter are crafted from carbon fiber, reducing overall weight without compromising structural integrity.

This meticulous approach extends beyond raw performance. The aesthetic qualities of carbon fiber are integrated into the car's design, giving the DBS Superleggera a visual edge that signals both power and elegance. Each panel undergoes extensive testing to ensure it meets Aston Martin's stringent quality standards, resulting in a car that is as resilient as it is visually striking.

Looking ahead, the DBS Superleggera represents a bridge to Aston Martin's hybrid and electrification strategy. While the current V12 engine delivers raw, naturally aspirated emotion, future iterations of the DBS are expected to integrate hybrid technology, combining electric torque with combustion power to enhance performance and efficiency.

Hybridization allows for instantaneous torque delivery, improving responsiveness in urban environments and providing greater flexibility in performance driving scenarios. It also signals Aston Martin's dedication to sustainability, ensuring the brand remains competitive in a world increasingly conscious of emissions and environmental impact.

The DBS Superleggera's engineering philosophy makes it an ideal platform for future innovation. Lightweight construction, advanced suspension, and active aerodynamics create a foundation that will allow hybrid powertrains to be integrated without compromising the car's dynamics, maintaining the thrilling driving experience that defines Aston Martin.

Beyond its grand touring prowess, the DBS Superleggera is engineered for precision on the track. Advanced traction and stability systems work seamlessly with the car's lightweight chassis, ensuring that the enormous power of the V12 is delivered in a controllable and predictable manner. Drivers can experience the full performance potential without compromise.

Every element of the car, from the brake system to the steering feel, has been optimized for responsiveness. Carbon-ceramic brakes provide exceptional stopping power, while the steering system delivers feedback with clarity and immediacy, creating a sense of connection between the driver and the road that few other vehicles can replicate.

The DBS Superleggera excels in balancing raw performance with driver confidence. Even when pushed to the limit, the car maintains composure, rewarding skilled drivers with an engaging, immersive, and exhilarating experience.

Aston Martin's commitment to exclusivity is fully realized in the DBS Superleggera. Through the Q by Aston Martin personalization program, owners can tailor every aspect of their vehicle, from interior materials to exterior finishes. Hand-stitched leathers, unique paint shades, and bespoke interior trims ensure that each DBS Superleggera is as individual as its owner.

This dedication to personalization aligns with Aston Martin's philosophy of creating not just cars but statements of personal taste and style. Limited-edition models further emphasize rarity, offering unique design touches and performance features that make each vehicle a collectible masterpiece.

The DBS Superleggera stands as both a culmination of Aston Martin's legacy and a benchmark for future models. It embodies the brand's dedication to combining cutting-edge engineering, timeless design, and unmatched craftsmanship. Each element, from its aerodynamic body to its luxurious interior, reinforces Aston Martin's philosophy of creating cars that inspire emotion while delivering performance.

As the automotive industry evolves, the DBS Superleggera will continue to serve as a model for integrating hybrid technology, advanced materials, and bespoke craftsmanship into a cohesive, world-class grand tourer. Its influence on future Aston Martin models will be profound, shaping everything from powertrain development to design language.

The DBS Superleggera is the embodiment of Aston Martin's vision of the ultimate grand tourer. With a V12 engine that thrills, a chassis that responds with precision, and a cabin that envelops drivers in luxury, it delivers a complete driving experience that few vehicles can match.

This car is more than a sum of parts; it is a statement of what is possible when performance, innovation, and craftsmanship converge. The DBS Superleggera exemplifies Aston Martin's ability to create vehicles that are timeless, exhilarating, and utterly luxurious—a benchmark in the world of grand tourers that will inspire admiration and desire for years to come.

Aston Martin has always understood that beauty and speed are inseparable, and the DBS Superleggera exemplifies this philosophy through its aerodynamic mastery. Every curve, vent, and surface has been meticulously designed to reduce drag, increase downforce, and optimize airflow. The car's active aerodynamic features, including its rear spoiler and front splitter, adjust dynamically at high speeds, improving grip and stability while maintaining the elegance of its silhouette.

The sculpted body panels not only enhance performance but also serve as a visual statement. Air is channeled precisely to cool critical components such as brakes and the twin-turbo V12 engine, ensuring optimal performance even during the most demanding driving scenarios. The carbon-fiber rear diffuser, integrated seamlessly with the rear bumper, directs air flow to improve traction and enhance the signature DBS Superleggera exhaust note.

Even minor details, such as the aerodynamic side skirts and hood vents, have been engineered with precision. These elements reduce lift and drag while adding a dynamic visual flair, demonstrating Aston Martin's ability to fuse form and function in a way that few other manufacturers can match.

While its powertrain is a marvel of mechanical engineering, the DBS Superleggera also integrates advanced technology to elevate the driving experience. Aston Martin has created a cockpit that connects driver and machine seamlessly, ensuring that every decision behind the wheel is informed, precise, and intuitive.

The digital instrument cluster provides a wealth of information, from speed and gear selection to adaptive suspension feedback and performance telemetry. Drivers can configure displays to highlight essential data while maintaining a clean, uncluttered aesthetic. Touchscreen infotainment, voice-activated controls, and smartphone integration ensure that the car meets modern expectations for connectivity without detracting from the driving experience.

Adaptive driving modes allow the DBS Superleggera to transform according to the driver's preferences. "GT" mode emphasizes comfort and composure, ideal for long-distance cruising, while "Sport" and "Sport Plus" modes sharpen throttle response, suspension stiffness, and steering feel. The car's electronic systems work harmoniously with the mechanical components, creating a vehicle that feels alive, responsive, and exhilarating.

For those who seek the ultimate in driving engagement, the DBS Superleggera excels on the track as well as the open road. The V12 engine delivers raw, unfiltered power, but it is the precision of the chassis and steering that ensures this power is always controllable. Carbon-ceramic brakes provide exceptional stopping power, capable of repeated high-speed deceleration without fade, while adaptive dampers maintain composure through corners, bumps, and high-speed straights.

The balance between weight, aerodynamics, and power delivery creates a driving experience that feels natural and intuitive. The DBS Superleggera responds instantly to steering inputs, with minimal body roll and razor-sharp agility. The tactile sensation of the steering wheel, the responsiveness of the throttle, and the crisp feedback of the brakes combine to create an immersive driving experience. Each journey—whether on a winding mountain pass or a racetrack—is a testament to Aston Martin's dedication to performance engineering.

Despite its ferocious performance, the DBS Superleggera never sacrifices comfort or luxury. Hand-stitched leather, polished carbon fiber, and carefully selected metals create a cabin environment that rivals the finest private lounges. The seats provide supreme support, while materials and finishes are chosen for both aesthetic beauty and tactile perfection.

Sound engineering is another hallmark of the interior experience. The exhaust note is tuned to resonate with emotion, while cabin acoustics are designed to create a rich and satisfying auditory environment. Whether accelerating on an open road or cruising through a city, every sound has purpose, reinforcing the sense that the DBS Superleggera is alive, responsive, and designed to excite every sense.

The DBS Superleggera is a canvas for Aston Martin's personalization philosophy. Through the Q by Aston Martin program, each owner can specify virtually every element of the car, from exterior colors to interior materials and stitching. Rare leather hides, custom paint finishes, and unique design details allow the DBS Superleggera to become a one-of-a-kind masterpiece.

Limited editions take this exclusivity further. Special variants, produced in extremely low numbers, feature bespoke design touches, enhanced performance components, and collectible appeal. Owning a DBS Superleggera is more than acquiring a car; it is embracing a statement of personal taste, artistry, and automotive sophistication.

The DBS Superleggera is a living link between Aston Martin's illustrious past and its ambitious future. The "Superleggera" name reflects decades of innovation in lightweight construction, while the current model integrates state-of-the-art technology, hybrid-ready engineering, and modern luxury.

This balance of heritage and innovation ensures the DBS Superleggera remains relevant in a rapidly evolving automotive landscape. Every curve, every stitch, and every component honors the company's legacy of craftsmanship, while simultaneously pushing the boundaries of what a modern grand tourer can achieve.

Ultimately, the DBS Superleggera is more than a car; it is a holistic automotive experience. It embodies speed, luxury, precision, and artistry, offering a driving experience that is both visceral and refined. Every detail, from the V12 engine's thunderous roar to the polished leather of the cabin, has been designed to inspire, thrill, and captivate.

As a benchmark for Aston Martin, the DBS Superleggera will influence future models, guiding design, performance, and technological development. It is a grand tourer that redefines expectations, a vehicle that celebrates the brand's heritage while pointing decisively toward its future. For drivers and enthusiasts alike, the DBS Superleggera is the ultimate expression of Aston Martin's pursuit of perfection.

16: CHALLENGES AND TRIUMPHS

The story of Aston Martin's revival is incomplete without acknowledging the remarkable leadership of Ulrich Bez, whose tenure as CEO from 2000 to 2013 transformed a struggling marque into a formidable contender in the luxury automotive world. Bez's approach was methodical yet inspired, blending the respect for the company's storied heritage with the pressing demands of modern automotive engineering. His philosophy was simple yet profound: a true Aston Martin should never compromise its identity as a symbol of elegance, performance, and exclusivity, even as it embraced new technologies and markets.

Bez quickly recognized that Aston Martin's future lay in a careful combination of product innovation, design excellence, and operational stability. Drawing on his engineering background, he implemented systematic changes across production, supply chains, and quality control, ensuring that every vehicle leaving the Gaydon factory met the highest standards of craftsmanship. Simultaneously, he sought to reignite the emotional connection between the brand and its enthusiasts, positioning Aston Martin as not just a car manufacturer, but a maker of dreams and experiences.

A pivotal factor in Aston Martin's resurgence was the cultivation of strategic alliances, most notably the long-standing relationship with Ford Motor Company. While Aston Martin retained its independence in design and brand philosophy, Ford provided the capital infusion and technical support needed to stabilize operations and invest in the next generation of vehicles. This partnership allowed Aston Martin to modernize production techniques, adopt new engineering innovations, and create cars that rivaled competitors in reliability without sacrificing the bespoke qualities that defined the brand.

One of the key advantages of this collaboration was access to advanced powertrain technology, which enabled the creation of engines that were both powerful and efficient. Bez leveraged these resources to develop vehicles that honored Aston Martin's heritage of performance while meeting contemporary expectations for speed, safety, and environmental responsibility. This combination of financial backing, technical know-how, and strategic vision set the stage for a revival that would astonish enthusiasts and critics alike.

The launch of the DB7 in 1994 marked a turning point not only for Aston Martin but for the luxury sports car market as a whole. Designed by Ian Callum, the DB7 was an exquisite blend of form and function, signaling a return to classic Aston Martin aesthetics while embracing modern engineering standards. It featured a six-cylinder engine that balanced performance with drivability, making it accessible to a wider audience without diluting the brand's aura of exclusivity.

The DB7's success was immediate and profound. Sales figures exceeded expectations, and the model became a touchstone for the revitalization of Aston Martin's reputation. Beyond its commercial success, the DB7 reinvigorated the design language of the company, establishing visual cues—elegant curves, signature grilles, and refined interiors—that would become hallmarks of subsequent models. Its introduction proved that a commitment to both engineering excellence and design integrity could restore confidence in a brand previously seen as teetering on the brink of decline.

Following the triumph of the DB7, Aston Martin continued its resurgence with the 2004 introduction of the DB9—a model that would define a new era for the brand. Equipped with a 6.0-liter V12 engine and an all-aluminum chassis, the DB9 combined blistering performance with structural sophistication. Its design, sleek yet assertive, echoed Aston Martin's heritage while offering aerodynamic efficiency suited for modern performance standards.

The DB9 was more than just a car; it was a statement of intent. It proved that Aston Martin could innovate without betraying its DNA, marrying hand-crafted luxury with cutting-edge technology. Inside, the cabin exemplified meticulous attention to detail, from the choice of leathers and veneers to the precision of the stitching and assembly. The car's driving dynamics were equally remarkable, offering a balance of power, agility, and comfort that appealed to discerning drivers around the globe.

The DB9 also marked Aston Martin's commitment to consistency in production quality. Every car was hand-built, ensuring that each model maintained a standard of craftsmanship that few competitors could match. Through the

DB9, Aston Martin reaffirmed its position as a luxury sports car manufacturer capable of producing vehicles that could compete on the world stage both technically and aesthetically.

The turn of the millennium was not without its challenges. Global economic instability, evolving emissions regulations, and shifts in consumer preferences demanded that Aston Martin adapt quickly. Yet, through strategic leadership and visionary product planning, the company navigated these obstacles with remarkable dexterity. By introducing vehicles like the DB7 and DB9, and later the V8 Vantage and DBS Superleggera, Aston Martin not only weathered market pressures but also carved a niche as a maker of high-performance luxury cars that remain aspirational symbols across continents.

Aston Martin's resurgence is also a testament to brand resilience. Despite the turbulence of the 1970s and 1980s and the uncertainties of the early 2000s, the company retained its core identity. Each challenge became an opportunity for innovation and refinement, enabling Aston Martin to emerge stronger and more revered than ever. The lessons learned from past struggles—particularly the importance of financial stability, visionary leadership, and technical excellence—have been embedded in the company's DNA, guiding its strategies to this day.

Ultimately, Aston Martin's story of challenges and triumphs is a celebration of persistence, vision, and unwavering commitment to quality. The struggles of the 1970s and 1980s, while daunting, paved the way for reinvention, illustrating that even the most storied brands must evolve to survive. Under Ulrich Bez's leadership, Aston Martin embraced both its heritage and the demands of the modern automotive world, creating a legacy that honors tradition while pushing boundaries.

Today, Aston Martin stands as a symbol of perseverance and excellence. Its vehicles continue to blend luxury, performance, and innovation, capturing the imaginations of enthusiasts worldwide. From the rebirth marked by the DB7 to the modern triumphs of the DB9, Vantage, and DBS Superleggera, Aston Martin has demonstrated that true greatness emerges not from the absence of challenges, but from the ability to overcome them with vision, craftsmanship, and passion.

As the company looks to the future—exploring electric powertrains, hybrid models, and advanced technologies—it does so with a foundation forged through decades of struggle, triumph, and unwavering commitment to excellence. Aston Martin's story serves as both an inspiration and a benchmark for luxury automotive brands, proving that resilience, innovation, and a respect for heritage can create timeless success.

Even as Aston Martin began regaining stability under Ulrich Bez, the road to full revival was not without further hurdles. The luxury sports car market remained fiercely competitive, with established European rivals like Ferrari, Porsche, and Bentley constantly innovating. Aston Martin's relatively small production scale, once a point of charm, now posed logistical and financial challenges. Meeting rising global demand while maintaining the handcrafted quality that defined the brand required careful balancing of resources, production planning, and technological integration.

Bez and his team approached these challenges strategically. Every decision, from introducing new models to refining supply chains, was guided by the principle that Aston Martin must remain exclusive yet aspirational. Partnerships with suppliers were strengthened, ensuring access to cutting-edge materials and components, while investment in in-house craftsmanship preserved the artisanal touch that distinguished every Aston Martin. It was a delicate dance: scaling operations without diluting the luxury experience, modernizing production while retaining hand-finished perfection.

The revival was not merely about survival; it was about reasserting Aston Martin's identity as a maker of world-class luxury performance cars. Models such as the V8 Vantage, introduced in 2005, became emblematic of the brand's resurgence. It offered a potent combination of raw power, agility, and style in a compact grand tourer package. The V8 Vantage was designed to appeal to both traditional enthusiasts and a younger demographic seeking performance with prestige, proving Aston Martin could evolve without sacrificing its core DNA.

Meanwhile, flagship models like the DB9 continued to define Aston Martin's aspirational appeal. With advanced aluminum architecture, refined aerodynamics, and luxurious interiors, the DB9 exemplified how the brand could merge performance with elegance. These vehicles did more than restore revenue—they reignited global enthusiasm for Aston Martin, reaffirming the brand's promise of handcrafted excellence combined with thrilling driving dynamics.

Aston Martin's comeback also relied on an ambitious global strategy. Under Bez, the company expanded its presence in key markets, including the United States, China, and the Middle East, regions where demand for high-performance luxury vehicles was growing. Aston Martin carefully curated its dealerships, ensuring each location reflected the brand's prestige and offered a luxury experience consistent with the company's ethos.

Marketing initiatives emphasized heritage and craftsmanship, highlighting the artisanal nature of Aston Martin vehicles and their association with elegance, sophistication, and British prestige. Collaborations with films, including the iconic James Bond franchise, reinforced Aston Martin's image as a symbol of style, sophistication, and adventure. This blend of product excellence and cultural visibility allowed Aston Martin to strengthen its global reputation while remaining true to its exclusive brand positioning.

While tradition remained central to Aston Martin's identity, innovation was equally crucial. The company embraced cutting-edge engineering, integrating advanced powertrains, chassis dynamics, and lightweight materials into new models. The DBS Superleggera, V12 Vantage, and DB11 exemplified this marriage of tradition and technology. Carbon fiber body components, advanced suspension systems, and turbocharged engines elevated performance while retaining the meticulous attention to detail expected in an Aston Martin.

At the same time, artisans continued hand-finishing every vehicle, ensuring the brand's hallmark craftsmanship was preserved. Leather was hand-selected and stitched, wood veneers polished to perfection, and metal accents meticulously applied. This balance of high-tech engineering and traditional craftsmanship became Aston Martin's defining formula for success, offering vehicles that were both technologically advanced and aesthetically timeless.

Aston Martin's journey from near collapse to resurgence illustrates the resilience and adaptability of a company committed to its values. By combining visionary leadership, strategic partnerships, and an unwavering commitment to quality, the brand overcame decades of uncertainty. It not only survived the turbulent 1970s and 1980s but emerged stronger, more cohesive, and globally recognized as a pinnacle of automotive luxury.

The lessons of this period continue to inform Aston Martin's approach today. Leadership emphasizes sustainable growth, innovation that respects brand heritage, and maintaining the exclusivity that makes every Aston Martin a collector's dream. Future models, including electrified and hybrid vehicles, are being developed with the same attention to detail and performance standards, ensuring the company evolves responsibly without compromising its identity.

In the end, Aston Martin's story is one of perseverance and triumph. The challenges of economic downturns, management instability, and technological evolution could have undone the brand, yet careful leadership and a steadfast commitment to quality ensured survival. From the DB7 to the DBS Superleggera, every model represents a milestone in Aston Martin's journey, reflecting a delicate balance between innovation, luxury, and performance.

Aston Martin's resurgence demonstrates that even the most storied brands must navigate adversity with vision, adaptability, and passion. Today, the company stands as a testament to the enduring power of craftsmanship, performance, and design excellence. Each vehicle is not merely a car—it is a symbol of triumph over adversity, a celebration of heritage, and a promise of future innovation.

The challenges of the past have shaped Aston Martin into a brand that can weather economic storms, technological shifts, and changing market trends. Its triumph lies not only in financial recovery but in reaffirming its position as a global icon of luxury, craftsmanship, and automotive excellence. As Aston Martin looks toward the future, it does so with the confidence that the lessons of its past will guide it in creating vehicles that continue to inspire, exhilarate, and captivate enthusiasts around the world.

17: SUSTAINABILITY AND THE ELECTRIC FUTURE

Aston Martin, long celebrated as the embodiment of luxury, performance, and elegance, stands today at a crossroads. The automotive world is shifting, guided by an urgent need for sustainability, cleaner energy, and environmental stewardship. Aston Martin recognizes that its legacy of high-performance vehicles must evolve in harmony with a planet-conscious future. The challenge is immense: how does a brand synonymous with roaring V12 engines, hand-stitched leather, and bespoke craftsmanship embrace sustainability without compromising the soul of what makes it Aston Martin?

The company's journey toward environmental responsibility is not a reactionary measure; it is a proactive commitment to redefine luxury in the modern era. Aston Martin is blending its rich heritage with green innovation, balancing tradition and progress. It is a brand that understands luxury is not only in speed or comfort but in the responsibility it bears toward the environment.

A sustainable future for Aston Martin begins with clarity of purpose. Recognizing the global push to reduce carbon emissions and the increasing scrutiny of automotive environmental impact, the brand has laid out a roadmap that intertwines regulatory compliance with visionary innovation. Its approach is multifaceted: reducing carbon emissions, implementing hybrid and electric technologies, and exploring renewable energy solutions across manufacturing processes.

Key to this strategy is the integration of hybrid and electric drivetrains across Aston Martin's lineup. While high-performance combustion engines remain central to the brand's identity, hybridization and electrification represent the bridge to a cleaner, sustainable era. Aston Martin sees electric mobility not merely as an alternative but as an opportunity to redefine what a luxury sports car can be in the twenty-first century. Every step in this roadmap is carefully measured, ensuring that performance, craftsmanship, and heritage are never compromised.

The company's long-term vision includes substantial reductions in fleet emissions, greener production facilities, and innovations in materials that align with environmental sustainability. These initiatives are designed not only to meet regulatory requirements but to cement Aston Martin as a global leader in luxury, performance-driven sustainability.

In 2018, Aston Martin took a bold step into the electric future by resurrecting its Lagonda marque as a fully electric brand. Lagonda, historically known for British innovation and engineering excellence since 1906, now serves as Aston Martin's laboratory for sustainable luxury. This revival is far more than branding; it is the company's statement that high-end performance and environmental consciousness can coexist.

Lagonda represents a reimagining of what a luxury vehicle can be in an electric age. The brand is tasked with combining the soul of Aston Martin's engineering with cutting-edge EV technology. From aerodynamics to materials, Lagonda cars are designed to minimize environmental impact while maximizing comfort, style, and performance. This approach acknowledges that sustainability must be integral to the luxury experience, rather than an afterthought.

The 2018 Lagonda Vision Concept exemplifies this philosophy. Sleek, futuristic, and unmistakably luxurious, it sets a standard for environmentally conscious automotive design. The Vision Concept emphasizes a carbon-neutral manufacturing process, showcasing Aston Martin's commitment to sustainable materials, energy-efficient production, and lightweight, high-performance components.

Beyond electrification, Aston Martin is pioneering green technologies across all facets of production. The company has publicly committed to carbon-neutral manufacturing, ensuring that each vehicle is produced with the smallest environmental footprint possible. Facilities at Gaydon, England, now incorporate renewable energy solutions, waste minimization strategies, and energy-efficient processes that align with this vision.

Materials selection also plays a critical role. Recyclable metals, responsibly sourced leathers, and bio-based composites are increasingly integrated into Aston Martin's vehicles. The focus is not just on reducing emissions but on creating products that demonstrate sustainability from the factory floor to the finished car.

Hybrid powertrains further support this vision, allowing Aston Martin vehicles to combine the thrills of traditional engines with the efficiency of electric motors. Hybrid models like the DBX707 SUV deliver the brand's signature power while reducing emissions, proving that environmental responsibility and driving exhilaration can coexist.

Hybrid powertrains are more than transitional technologies; for Aston Martin, they are a cornerstone of sustainable performance. The company faces a delicate balancing act: maintaining the visceral, emotive driving experiences for which it is famed while meeting increasingly strict environmental standards.

By integrating hybrid systems into high-performance vehicles, Aston Martin demonstrates that electrification need not sacrifice the brand's core identity. Hybrid engines offer quiet, efficient operation at lower speeds while unleashing full combustion-powered performance when demanded. For drivers, this means a seamless blend of sustainability and exhilaration.

Furthermore, Aston Martin is actively exploring advanced hybrid designs and hydrogen fuel cells for potential future applications. Hydrogen, with water as its only byproduct, represents a potential breakthrough in zero-emission mobility. Though still in development, these technologies signal Aston Martin's commitment to long-term innovation and environmental leadership.

The resurrection of the Lagonda brand marks a bold statement in Aston Martin's sustainability journey. No longer merely a historic nameplate, Lagonda represents the pinnacle of electrified luxury. Every aspect of its design—from aerodynamics to interiors, materials, and even manufacturing processes—has been reimagined with environmental consciousness at its core. Aston Martin envisions Lagonda as a series of fully electric vehicles (EVs) that rival the traditional combustion-powered supercars in both performance and desirability.

The futuristic designs, particularly evident in the Lagonda Vision Concept, showcase long, sweeping lines, minimalistic interiors, and advanced driver-centric technology. The cabins are meticulously crafted, combining sustainable materials like plant-based leathers, recycled composites, and responsibly sourced metals. Every surface communicates luxury, yet with a commitment to environmental responsibility. This is Aston Martin's philosophy: sustainability does not compromise elegance or performance; it elevates them.

Lagonda is also pioneering innovations in electric powertrain architecture. Aston Martin engineers are developing systems that maximize efficiency while maintaining the instantaneous torque and responsive handling expected of a British grand tourer. The focus is not only on environmental impact but on delivering a driving experience that embodies the brand's storied heritage of exhilaration, precision, and refinement.

Sustainability at Aston Martin extends beyond electrification—it permeates every stage of the production process. The company has committed to carbon-neutral manufacturing, investing in renewable energy, energy-efficient production lines, and waste reduction initiatives. At its Gaydon headquarters, solar and wind power integration, combined with state-of-the-art energy management systems, ensures minimal carbon output during the production of each vehicle.

Moreover, Aston Martin has embraced the recycling and repurposing of materials wherever possible. Older aluminum panels are melted and reused, and composite materials are carefully processed to reduce landfill waste. By combining traditional craftsmanship with modern sustainability techniques, the company ensures that each car is both an artisanal masterpiece and an environmentally responsible product.

This approach reflects a broader trend in the luxury automotive industry: the understanding that performance and sustainability are not mutually exclusive. Aston Martin is proving that eco-conscious production can enhance quality, rather than diminish it, while reinforcing the brand's reputation for unparalleled craftsmanship.

Hybrid and electric powertrains are central to Aston Martin's strategy, bridging the gap between traditional petrol engines and a fully electric future. Models such as the DBX707 showcase how hybridization can enhance both performance and environmental efficiency. The combination of a turbocharged V8 engine and electric motors delivers instant torque, smooth acceleration, and reduced emissions—a seamless marriage of power and responsibility.

Looking further ahead, Aston Martin is exploring advanced battery technology and modular electric architectures to maximize range and efficiency. The focus is on delivering performance parity with combustion engines while meeting stringent global environmental regulations. Battery management systems, regenerative braking, and lightweight materials all contribute to a driving experience that is unmistakably Aston Martin: visceral, precise, and thrilling.

Hydrogen fuel cells are also under consideration for future models, offering the potential for zero-emission operation while maintaining range and power. Aston Martin's research in this area emphasizes innovation and foresight, ensuring that the brand remains at the cutting edge of automotive technology as the market shifts toward electrification.

Sustainability is not limited to propulsion systems. Aston Martin is rethinking materials to reduce environmental impact without compromising the tactile and visual luxury expected by its clientele. Interiors increasingly feature plant-based leathers, natural fibers, recycled metals, and responsibly sourced exotic woods.

Every stitch, trim, and surface is selected with sustainability in mind, reflecting the brand's philosophy that luxury and responsibility can coexist. Handcraftsmanship remains at the core, but it is enhanced with eco-conscious innovations. Even paint systems are evolving, with water-based and low-VOC finishes replacing traditional solvents, reducing both factory emissions and environmental impact.

Aston Martin recognizes that technology is the key to achieving sustainability goals without sacrificing performance. Advanced aerodynamics, lightweight carbon fiber and aluminum composites, and next-generation battery systems allow the vehicles to remain light, agile, and powerful while significantly reducing energy consumption.

Connected vehicle technologies also contribute to efficiency. Intelligent energy management systems, adaptive regenerative braking, and route-optimization software maximize battery range and minimize waste. Aston Martin's approach demonstrates that sustainable technology can enhance performance rather than detract from it, preserving the brand's commitment to an exhilarating driving experience.

Transitioning a luxury brand with a storied history to a fully electric and sustainable future is no small feat. Challenges include developing high-performance batteries without compromising range, maintaining the traditional driving feel, and ensuring that new technologies align with the brand's heritage of elegance and craftsmanship.

Yet these challenges represent opportunities. By pioneering sustainable luxury vehicles, Aston Martin is setting new benchmarks in the industry. Electric and hybrid vehicles open the door to innovations in autonomous driving, enhanced safety features, and connectivity—all of which can be integrated without undermining the driving experience.

Aston Martin's vision extends beyond its product lineup. The brand aims to become a global leader in sustainable luxury manufacturing, demonstrating that environmental responsibility can coexist with elegance, performance, and exclusivity. By investing in electric technology, hybrid powertrains, and carbon-neutral production, Aston Martin is redefining the concept of a luxury car for the 21st century.

The company's strategy underscores a central belief: true luxury must embrace responsibility. The vehicles of the future will not only thrill their drivers but also respect the world they inhabit. Aston Martin's commitment ensures that its cars remain aspirational, timeless, and forward-looking, proving that high performance and sustainability are not only compatible—they are complementary.

Looking ahead, Aston Martin is poised to influence the global transition toward sustainable mobility. Its roadmap of electrification, green manufacturing, and hybrid integration serves as a model for other luxury automakers. By merging heritage with innovation, Aston Martin is ensuring that the future of performance vehicles is also a future of environmental stewardship.

In this journey, Aston Martin remains true to its core identity: performance, design, and luxury. Yet now, each roaring engine, each handcrafted interior, and each aerodynamic curve carries with it a commitment to a cleaner, greener world. Aston Martin's electric future is a vision where sustainability and exhilaration coexist, ensuring that the brand continues to lead not just in luxury performance but in responsible innovation.

18: THE GLOBAL ASTON MARTIN COMMUNITY

Aston Martin is far more than a manufacturer of luxury sports cars; it is a symbol, an emblem of elegance, performance, and an enduring sense of exclusivity. Over the course of more than a century, the brand has cultivated a loyal following of enthusiasts who see Aston Martin not merely as an automobile, but as a lifestyle, a statement, and an entry into a community of like-minded aficionados. These are individuals who are bound not only by their appreciation for high-performance engineering and aesthetic mastery, but by a shared reverence for tradition, craftsmanship, and innovation.

The Aston Martin community stretches across continents, uniting owners and fans from the bustling streets of Tokyo to the sun-kissed coastlines of California, from the historic roads of England to the sweeping highways of Australia. In every corner of the globe, Aston Martin owners form tight-knit circles, bonded by a passion for the marque that transcends the cars themselves. This is a community that thrives on shared experiences—whether attending a grand driving tour, participating in track events, or discussing the minute details of engineering excellence over a coffee in a local clubhouse.

At the heart of this community lies a unique duality: a respect for Aston Martin's heritage paired with an excitement for its modern evolution. Collectors cherish vintage models like the DB4 and the iconic DB5, while younger enthusiasts celebrate the dynamic performances of the Vantage, DBS Superleggera, and DBX. Together, this community embodies both the history of Aston Martin and its future, creating a network that is diverse, passionate, and uniting.

The Aston Martin brand is an amalgamation of artistry, engineering precision, and a storied racing pedigree. It is a company that, since its inception in 1913, has combined performance engineering with bespoke craftsmanship. Every vehicle rolling off the Gaydon production line is a testament to this philosophy, meticulously assembled by artisans whose skills have been honed over decades. From the leatherwork in the cabin to the precision-tuned engine, every component is carefully designed to evoke emotion and pride.

This attention to detail and commitment to quality resonates deeply within the community. Enthusiasts are drawn to Aston Martin not just for the thrill of speed or the elegance of design, but for the culture of excellence it represents. Owning an Aston Martin is a badge of membership in a tradition that values refinement, innovation, and exclusivity. The community is bound by these shared values, creating a sense of identity and belonging that is unique among automotive brands.

Whether it is through private club events, high-profile rallies, or online discussions, the Aston Martin community celebrates these qualities. There is an unspoken understanding among owners: the car is not merely a means of transportation—it is a canvas for artistry, a machine built to excite, and a legacy to uphold. In this way, the vehicles themselves become ambassadors for the culture of the brand, fostering pride and admiration among those who are fortunate enough to own or interact with them.

Among the most influential pillars of the global community is the **Aston Martin Owners Club (AMOC)**, established in 1935. Recognized as one of the oldest and most prestigious automotive clubs in the world, AMOC has played a central role in cultivating a sense of unity among Aston Martin enthusiasts. Membership spans the globe, and the club acts as both a social nexus and a repository of knowledge, guiding owners through restoration projects, technical questions, and lifestyle experiences tied to the marque.

AMOC's events are as varied as they are meticulously organized. Track days allow members to experience the sheer thrill of high-performance driving, while road rallies provide scenic journeys where camaraderie and automotive excellence merge. Social gatherings, often hosted at luxurious estates or renowned hotels, provide an opportunity to celebrate the aesthetic and mechanical achievements of Aston Martin models, from the vintage DB2 to the latest DBX707. Through these events, members develop lifelong friendships rooted in mutual admiration for the brand.

Equally significant is the role AMOC plays in preserving Aston Martin's heritage. Classic models, such as the DB5 and the Lagonda, require meticulous restoration and care, and AMOC provides a forum for sharing knowledge, techniques, and resources. By fostering these efforts, the club ensures that Aston Martin's rich history remains vibrant and accessible, while simultaneously inspiring future generations of enthusiasts to maintain the brand's legacy.

In the United States, the Aston Martin Owners Club North America (AMOCNA) exemplifies the reach and vitality of the global Aston Martin community. Members from coast to coast engage in road rallies through iconic American landscapes, from the twisty Blue Ridge Parkway to the desert expanses of Nevada. These events provide a thrilling intersection of automotive performance, scenic exploration, and social bonding.

Across Europe, local chapters organize weekend drives along Italy's Amalfi Coast, the Swiss Alps, or the French Riviera, emphasizing not only performance but elegance in every mile traveled. Meanwhile, in Asia, Aston Martin clubs in Japan, China, and the Middle East have developed vibrant communities of collectors and enthusiasts who share the brand's ethos and commitment to luxury. Each chapter, though geographically diverse, shares a commitment to celebrating Aston Martin's engineering brilliance and iconic design language.

These clubs are more than social organizations; they are extensions of the brand itself. They serve as incubators for discussion, education, and mentorship, where experienced owners guide newcomers in understanding the complexities of maintaining, driving, and appreciating an Aston Martin. This fosters a multigenerational culture, where the brand's legacy is continually passed down, sustaining its allure for decades to come.

For those fortunate enough to participate, Aston Martin events provide experiences that are unparalleled in the automotive world. The **Aston Martin Festival** is one such event, a gathering that unites owners, collectors, and fans for a weekend of immersive experiences. From high-speed track sessions to road tours and concours d'elegance competitions, the festival celebrates both the performance and artistry of Aston Martin vehicles.

Similarly, the **Aston Martin Racing Experience** caters to motorsport enthusiasts, allowing participants to drive race-spec vehicles on professional circuits under the guidance of seasoned drivers. This event creates a visceral connection to Aston Martin's racing heritage, allowing fans to experience the brand's performance pedigree firsthand. For many, it is a transformative experience—an opportunity to feel the synergy between precision engineering and human skill.

Concours d'elegance competitions, held at prestigious locations like Pebble Beach, California, or Villa d'Este in Italy, highlight Aston Martin's historical and artistic significance. Here, classic models are judged for their craftsmanship, originality, and preservation. For enthusiasts, these events are not merely competitions—they are celebrations of artistry, heritage, and the meticulous care required to maintain a brand's legacy.

As the world becomes increasingly connected, online communities have emerged as another cornerstone of the Aston Martin network. Social media platforms such as Instagram, Facebook, and Twitter offer a virtual space for enthusiasts to share images, videos, and stories. Owners showcase their vehicles, post updates on restoration projects, and provide insights into the latest models, fostering engagement among global fans.

Online forums such as **"Aston Martin Life"** and **"Aston Martin Chat"** allow enthusiasts to exchange technical knowledge, discuss maintenance strategies, and debate the merits of specific models. These platforms are invaluable for connecting newcomers with seasoned owners, creating a culture of mentorship and knowledge-sharing. The global reach of digital platforms ensures that Aston Martin's community is no longer limited by geography, enabling passionate fans from different continents to connect, collaborate, and celebrate the brand together.

The brand itself actively participates in these digital arenas, offering followers behind-the-scenes content, first looks at new releases, and interactive engagement. This level of accessibility strengthens the bond between Aston Martin and its audience, transforming passive admiration into an active and engaged community.

Aston Martin is a brand that has long been intertwined with celebrity culture, which has only strengthened the global community. The allure of driving an Aston Martin is not confined to car enthusiasts alone; it extends to those in film, music, and business who recognize the marque as a symbol of taste, status, and success. From Hollywood stars to international entrepreneurs, ownership of an Aston Martin carries with it an implicit statement of refinement and achievement.

Iconic appearances of Aston Martin in films—most notably the James Bond franchise—have contributed immensely to the brand's mystique. The DB5, first featured in *Goldfinger* (1964), remains one of the most celebrated vehicles in cinematic history, creating a cultural connection that endures across generations. These appearances foster a sense of pride and excitement within the community, as fans are reminded that their cars are not merely machines but icons of style, innovation, and daring.

Beyond the silver screen, celebrities frequently share their Aston Martins on social media, giving followers glimpses of the lifestyle associated with the brand. These public endorsements reinforce Aston Martin's identity as a marque that merges performance with prestige, and they encourage fan participation in global discussions, events, and online communities. The brand's celebrity connections contribute to its continued relevance and appeal, cultivating aspirational admiration among fans worldwide.

Aston Martin's influence transcends automobiles—it shapes art, fashion, and design culture globally. The marque's aesthetic language has inspired designers, architects, and creatives seeking to emulate the harmony between form and function that is inherent in every model. Sculptors study the fluidity of the DB11's bodywork; fashion designers take cues from the elegance of the leather interiors; even luxury yacht designers reference Aston Martin's blend of technology and sophistication.

The community thrives on these cross-disciplinary inspirations, fostering discussions that range from mechanical performance to design philosophy. Owners share insights about how Aston Martin principles can inform other creative ventures, creating a dialogue between car enthusiasts and design enthusiasts alike. This cultural reach strengthens the sense of global community, proving that Aston Martin is not just about vehicles, but a symbol of refinement that resonates in multiple spheres of influence.

Aston Martin owners do not merely drive; they live a lifestyle enriched by their vehicles. The brand's community is tightly linked to experiences that integrate travel, culture, and luxury. From organized road trips through Italy's scenic Amalfi Coast to high-speed track events in Germany's Nürburgring, owners engage in journeys that combine performance driving with cultural immersion.

These lifestyle experiences are amplified by Aston Martin's partnerships with luxury hotels, resorts, and private clubs, which provide exclusive access to venues tailored for owners. Weekend escapes often include curated experiences: fine dining, private tours, and interactive workshops on automotive craftsmanship. This blending of driving and lifestyle fosters deeper connections within the community, transforming ownership into a social and cultural journey.

Additionally, Aston Martin frequently collaborates with luxury brands to enhance these experiences, offering limited-edition accessories, custom-tailored apparel, and bespoke travel packages. These collaborations reinforce the idea that being part of the Aston Martin family is about more than owning a car; it's about embracing a world of sophistication, performance, and curated lifestyle experiences.

While in-person experiences are crucial, the digital era has revolutionized how the Aston Martin community interacts. Social media, online forums, and virtual events have created new avenues for engagement, connecting fans across borders in ways previously impossible. Enthusiasts can now share restoration projects, discuss technical specifications, or livestream their participation in track events with a global audience.

Instagram and Facebook provide visually immersive platforms for owners to showcase their vehicles, sharing moments from road trips, car shows, or rare models. Twitter and YouTube offer spaces for discussion, analysis, and engagement, where fans debate engine performance, aerodynamics, and design nuances. Even platforms like TikTok have seen a rise in Aston Martin-related content, reaching younger audiences and introducing a new generation to the brand's legacy.

Online communities are not merely passive spaces; they serve as catalysts for action. Coordinated virtual meetups, digital rallies, and live-streamed panels allow enthusiasts to feel part of a global family, regardless of physical location. This digital interconnectivity strengthens relationships within the community and ensures that Aston Martin's ethos is shared, celebrated, and continually expanded on a worldwide scale.

The Aston Martin community also thrives on its motorsport heritage. Aston Martin Racing, the brand's competitive division, has produced numerous victories in endurance racing and GT championships. For enthusiasts, these achievements are more than statistics—they are a source of inspiration and a tangible connection to the brand's performance philosophy.

The **Aston Martin Racing Experience**, a staple for many fans, offers a unique opportunity to experience the thrill of racing firsthand. Participants receive coaching from professional drivers, learning to handle high-performance machines on famous circuits. These events provide both technical education and exhilarating entertainment, allowing the community to experience what it truly means to be behind the wheel of an Aston Martin in its racing element.

Engagement with motorsport further strengthens the bonds of the global community. Fans share experiences, swap insights on driving techniques, and celebrate victories collectively, creating a sense of shared purpose and pride. Motorsport is a powerful unifier, connecting enthusiasts across continents through the universal language of speed, precision, and passion.

A significant aspect of the global community is the reverence for Aston Martin's classic models. Collectors invest in maintaining, restoring, and showcasing historic cars such as the DB4, DB5, and Lagonda. These vehicles are not simply valuable assets; they are living pieces of automotive history, preserving the artistry, engineering, and craftsmanship of previous generations.

Concours d'elegance events serve as gathering points for collectors and enthusiasts. These meticulously curated shows celebrate restoration excellence, originality, and design aesthetics. Participation in such events is a source of pride for owners, while spectators gain insight into the meticulous care required to maintain Aston Martin's legacy. Classic car communities reinforce the idea that ownership is a responsibility as much as it is a privilege, emphasizing respect for the marque's heritage.

Aston Martin's influence extends beyond enthusiasts and collectors. Its role in popular culture, from James Bond films to luxury lifestyle collaborations, has cemented the brand as a global symbol of sophistication. Music videos, fashion campaigns, and international media frequently feature Aston

Martin, enhancing its visibility and embedding the marque into the collective imagination.

Through these intersections of culture, performance, and luxury, the global community is continually energized. Fans are drawn not only by the cars themselves but by the experiences, stories, and lifestyles that the brand represents. Aston Martin transcends the physicality of automobiles, becoming a cultural touchstone that connects owners, enthusiasts, and admirers worldwide.

The global Aston Martin community is diverse, dynamic, and deeply committed. From private owners' clubs and racing enthusiasts to digital communities and celebrity patrons, the network of individuals bound to the brand is extraordinary in both scale and passion. This family is united not merely by admiration for automobiles, but by a shared appreciation for performance, design, heritage, and lifestyle.

As the brand continues to innovate, expand its product offerings, and embrace sustainability, the community will evolve alongside it. New models, advanced technologies, and lifestyle experiences will attract a new generation of enthusiasts while remaining true to the brand's core values. Aston Martin's community, therefore, is not static; it is a living, evolving organism, reflecting both the history and future of this iconic marque.

The Aston Martin family stands as a testament to the enduring power of brand loyalty, passion, and shared experience. For members, owning or admiring an Aston Martin is more than a hobby—it is an entry into a global network defined by elegance, performance, and camaraderie. It is a community that celebrates automotive artistry, values craftsmanship, and inspires a lifelong connection to a brand that continues to captivate hearts across the globe.

Aston Martin is more than an automotive manufacturer; it is a cultural phenomenon and a unifying force for enthusiasts worldwide. Through clubs, events, digital communities, motorsport heritage, and lifestyle integration, the global Aston Martin family embodies the brand's core values of **performance, luxury, craftsmanship, innovation, and legacy**. The community ensures that Aston Martin remains not just a car, but a symbol—a worldwide network that continues to thrive, inspire, and grow.

19: ASTON MARTIN AND POPULAR CULTURE

Aston Martin is far more than an automotive manufacturer. It is a symbol—a rolling emblem of sophistication, opulence, and refined taste. Over the decades, the marque has evolved from an exclusive car company into a cultural icon, bridging the worlds of performance engineering and high society. Its vehicles, with their sleek lines, rumbling engines, and unmistakable presence, are instantly recognizable. Yet, the true power of Aston Martin lies not merely in its mechanical excellence but in the aura it evokes, an aura that permeates cinema, music, fashion, and art.

For enthusiasts and collectors, owning an Aston Martin is akin to possessing a piece of living history. The brand's global influence extends well beyond the automotive industry, shaping lifestyle, aspiration, and desire. It is a company that has successfully transformed its engineering mastery into a symbol of aspiration, elegance, and cultural significance.

No discussion of Aston Martin's impact on popular culture is complete without examining its cinematic legacy. The brand first achieved international notoriety in 1964 with the James Bond film *Goldfinger*, in which the Aston Martin DB5 became the ultimate spy vehicle. Equipped with machine guns, revolving license plates, and a bulletproof shield, the silver DB5 was not merely a car—it was a character in its own right. Its association with the suave, sophisticated, and lethal spy James Bond immortalized Aston Martin as a symbol of elegance, power, and innovation.

From that moment onward, Aston Martin became inseparable from cinematic glamour. The relationship between the British marque and the Bond franchise has endured for decades, with successive films featuring models such as the DBS Superleggera, V8 Vantage, and the timeless DB5 in modernized roles. Aston Martin's vehicles are not just props; they shape the narrative, enhance the story, and signify the elite status of the characters who drive them. Bond's Aston Martin is a testament to the brand's dual nature: a harmonious marriage of style and raw performance.

Beyond James Bond, Aston Martin has featured prominently in Hollywood films that emphasize wealth, sophistication, and intrigue. In *The Italian Job* (1969) and *Casino Royale* (2006), Aston Martins are more than vehicles—they are status symbols, extensions of the characters' identities. From elegant chases through city streets to breathtaking countryside pursuits, the presence of an Aston Martin instantly elevates the cinematic scene, providing audiences with a visual cue of exclusivity, precision, and prestige.

Aston Martin's cultural footprint extends far beyond film, resonating deeply within the music world. Rap, hip-hop, and pop artists frequently reference Aston Martin in their lyrics, using the vehicle as a metaphor for success, ambition, and high living. Jay-Z, Drake, and Kanye West, among others, have celebrated the marque in their songs and music videos, highlighting its association with affluence and luxury.

The allure is clear: Aston Martin is more than a car; it is a symbol of aspiration. Its elegant silhouette, roaring engine, and polished exterior convey messages of achievement and refinement. In music videos, the vehicle is often used as a visual shorthand for opulence—a backdrop that communicates wealth and influence without words. The car is elevated to an icon of lifestyle, complementing the music, lyrics, and persona of the artist.

The influence of Aston Martin also extends to the world of fashion, where design, elegance, and visual storytelling intersect. The vehicles are not merely showcased—they are styled into campaigns, photo shoots, and luxury advertisements. High-end designers frequently pair their products with Aston Martin vehicles, merging two worlds defined by craftsmanship, attention to detail, and aesthetic perfection.

For celebrities, stepping out of an Aston Martin is a statement. It conveys taste, discernment, and prestige. Whether gliding along a city boulevard or appearing at a gala, the car enhances the individual's presence. The curves, the chrome accents, the balanced proportions of an Aston Martin are as deliberate and intentional as the cut of a designer garment, creating a synergy between the automotive and fashion worlds. Aston Martin is more than a tool for travel; it is a statement of elegance, style, and elite identity.

Aston Martin vehicles are celebrated not only for performance but for their artistry. Each model is designed with an eye toward perfection, blending aesthetics with functionality in ways that resonate with artists, collectors, and enthusiasts. The brand has collaborated with sculptors, designers, and visual artists to create limited-edition models and art-inspired concepts, elevating the automobile to a medium of cultural expression.

These collaborations go beyond mere marketing—they reflect Aston Martin's commitment to blending mechanical prowess with visual and tactile beauty. Cars are often displayed at galleries, art exhibitions, and auctions, blurring the line between engineering and fine art. For collectors, owning an Aston Martin is akin to possessing a piece of contemporary artistry, a manifestation of human creativity and technical mastery.

Ownership of an Aston Martin conveys more than personal taste—it communicates status, influence, and cultural literacy. Celebrities from Hollywood to the sports world have been drawn to the marque, using the car as both a tool and a symbol. Its presence in social media posts, red carpet appearances, and lifestyle shoots perpetuates its aura of exclusivity.

The brand's alignment with high-profile figures strengthens its cultural relevance, making it aspirational to a global audience. Ownership is not just functional; it is performative, a way of signaling success, refinement, and access to elite networks. Aston Martin's status as a symbol of luxury living is reinforced daily by its connection to the world's most influential personalities.

Aston Martin's cultural influence remains strong in the modern age, with the brand continuing to inspire and captivate. From film to music, fashion, and art, the marque has become more than a manufacturer—it is a global symbol of elegance, luxury, and achievement.

As the automotive landscape evolves with electrification, hybrid technologies, and digital innovation, Aston Martin retains its identity as a symbol of sophistication and performance. The Lagonda EV series, alongside models like the DBS Superleggera, ensures that Aston Martin's heritage of excellence continues while embracing the future. The marque's position in popular culture is secure, cementing its status as not merely a car brand but as a statement of ambition, creativity, and taste that transcends generations.

Beyond films, music, and art, Aston Martin's cultural significance extends into the very fabric of global luxury living. The marque has become synonymous with aspirational lifestyles, from private clubs and exclusive events to high-end resorts and luxury experiences. Owning an Aston Martin is often paired with access to a broader world of sophistication—gourmet dining, yachting, fashion shows, and elite social circles. The brand has cultivated not just a product but a culture, where the car is a gateway to a lifestyle defined by excellence, refinement, and prestige.

Owners of Aston Martin vehicles frequently report that the experience transcends the road itself. Attending events, participating in road rallies, and joining exclusive gatherings allows them to engage with a community that shares their passion. This global lifestyle aspect of Aston Martin reinforces its status not merely as an automobile but as a social and cultural institution. The brand's reach extends from Europe to North America, Asia, and the Middle East, creating a network of enthusiasts bound by a shared admiration for performance, design, and heritage.

In the digital age, Aston Martin has leveraged social media platforms to solidify its cultural influence. Instagram, Twitter, and Facebook serve as windows into the brand's world, offering fans access to breathtaking visuals of the latest models, behind-the-scenes factory tours, and exclusive events. These platforms create a dynamic dialogue between the company and its audience, fostering engagement with both owners and aspirants alike.

Digital campaigns often highlight the artistry, performance, and craftsmanship inherent in each vehicle. Short films, cinematic advertisements, and interactive posts showcase the cars not only as machines but as expressions of luxury and innovation. Online forums, fan pages, and enthusiast groups further amplify this reach, allowing owners and admirers to share experiences, discuss restorations, and celebrate the marque. Through digital channels, Aston Martin transcends geography, cultivating a global fan base that spans continents and cultures.

Aston Martin enthusiasts are drawn together not only online but in real-world experiences that celebrate performance and luxury. Exclusive gatherings such as the Aston Martin Festival, track days, and concours d'elegance events create immersive environments where owners can showcase their cars, participate in high-speed driving experiences, and socialize with fellow enthusiasts.

The Aston Martin Racing Experience, for instance, allows owners to take the wheel of race-spec vehicles on professional circuits under expert guidance. These events combine exhilaration with education, offering insight into the engineering excellence behind the brand while reinforcing the sense of community and prestige. Competitions and rallies, often held across scenic routes worldwide, allow owners to experience their vehicles' capabilities while building connections rooted in shared passion.

Concours d'elegance events highlight the historical and artistic value of Aston Martins. From the classic DB5s to modern masterpieces, these showcases celebrate both design excellence and the heritage of automotive artistry. Owners often invest significant effort in restoring and maintaining vehicles to pristine condition, reinforcing the brand's reputation for craftsmanship while creating visually stunning spectacles that draw international attention.

Aston Martin's association with celebrities continues to amplify its cultural reach. High-profile figures from entertainment, sports, and business often choose Aston Martins as symbols of success, glamour, and status. Social media posts featuring these vehicles introduce the brand to millions, reinforcing its perception as an elite, desirable marque.

Hollywood stars, renowned musicians, and sports icons have all contributed to Aston Martin's aura, elevating its presence in lifestyle media and digital storytelling. Each celebrity appearance reinforces a narrative: owning an Aston Martin is an aspirational act, a visual manifestation of achievement and refined taste. This form of cultural endorsement strengthens the connection between the brand and its admirers, bridging the gap between automotive performance and symbolic prestige.

Aston Martin has masterfully integrated itself into media, collaborating with filmmakers, photographers, and content creators to craft narratives that emphasize the brand's elegance, power, and innovation. Advertisements and short films often depict the cars in cinematic sequences, highlighting performance capabilities alongside aesthetics. Storytelling plays a critical role in these campaigns, linking Aston Martin vehicles to themes of adventure, sophistication, and ingenuity.

This narrative-driven approach has allowed the brand to maintain cultural relevance, even as consumer tastes evolve. By focusing on aspirational storytelling, Aston Martin transcends the status of a product manufacturer, positioning itself as a lifestyle curator, a creative collaborator, and a symbol of global luxury living.

Through its enduring presence in films, music, fashion, art, and social media, Aston Martin has become a symbol whose influence extends far beyond the automotive industry. The brand represents more than speed or engineering—it embodies aspiration, refinement, and achievement.

Its cultural legacy continues to grow, ensuring that new generations recognize Aston Martin as a pinnacle of automotive artistry and lifestyle excellence. From iconic silver DB5s to futuristic Lagonda EVs, the marque bridges the gap between tradition and modernity, performance and sustainability, engineering mastery and artistic expression. Its continued relevance is a testament to careful brand stewardship, cultural engagement, and an unwavering commitment to excellence.

Aston Martin's position in popular culture is unparalleled among luxury automotive brands. Its influence permeates cinema, music, fashion, art, and lifestyle, creating a global network of admirers, owners, and enthusiasts. The cars themselves, while extraordinary in performance and design, serve as vessels for broader cultural significance, representing aspiration, achievement, and timeless style.

As the brand evolves toward electric mobility and sustainable luxury, it does so without sacrificing the elegance, craftsmanship, and performance that have cemented its cultural standing. Aston Martin is not merely a car—it is an icon, a cultural touchstone, and a statement of refined taste that continues to inspire the world. For generations past, present, and future, the marque remains a symbol of excellence, innovation, and the enduring allure of British automotive artistry.

20: THE FUTURE OF ASTON MARTIN

Aston Martin has always stood as a beacon of **luxury, innovation, engineering, elegance, and performance**. For over a century, the British marque has captured the imaginations of car enthusiasts, collectors, and even pop culture audiences worldwide. Its vehicles are more than cars—they are statements of aspiration, symbols of achievement, and mechanical works of art. Yet, as the 21st century continues to unfold, Aston Martin faces a new era defined not only by technological progress but also by global environmental and societal changes. The automotive world is evolving faster than ever, with electric vehicles, hybrid powertrains, autonomous technologies, and new global markets redefining the rules of the road. Aston Martin's challenge is to navigate these shifts while preserving the qualities that have made it iconic: **luxury, craftsmanship, performance, beauty, and exclusivity**.

The company has already begun charting a bold course into the future, emphasizing sustainable powertrains, cutting-edge design, and advanced connectivity. This new era is not simply about replacing engines or adding technology—it is about reimagining what a luxury vehicle can represent in an age of environmental consciousness and global innovation. The future of Aston Martin promises to be an amalgamation of its rich heritage and forward-thinking strategies, combining emotional design with scientific ingenuity.

The heart of Aston Martin has always been its engines, the throbbing lifeblood of every model. For decades, the brand has been celebrated for its **V12-powered** machines, which offered unparalleled sound, acceleration, and performance. However, the world is changing. Environmental awareness, government regulations, and consumer demand are shifting the focus from pure combustion power to hybrid and electric alternatives.

Aston Martin recognizes that the future lies in blending tradition with innovation. Its engineers are working tirelessly to design hybrid systems that combine the visceral excitement of combustion engines with the efficiency and sustainability of electric power. This hybrid approach allows Aston Martin to **retain its signature driving thrill** while meeting the expectations of a modern audience increasingly focused on environmental responsibility.

The company's Lagonda brand, resurrected as a fully electric luxury marque, is central to this vision. The **Lagonda Vision Concept**, unveiled in 2018, demonstrated that electric cars could embody both elegance and technical sophistication. With a sleek, futuristic exterior and zero-emissions powertrain, Lagonda represents a blueprint for the electric future of luxury automobiles. But Aston Martin's ambitions extend beyond just one model or brand—it aims to integrate electrification across its entire lineup, ensuring that every new Aston Martin blends sustainability with performance and artistry.

Creating an electric future is not merely about swapping engines; it requires a reimagining of the very essence of an

Aston Martin. Designers, engineers, and strategists are considering the interplay between power delivery, handling, weight distribution, and driver experience in ways never before required for traditional combustion vehicles. Each component, from chassis construction to aerodynamics, is being optimized for an electric-first world.

The roadmap also emphasizes collaboration. Aston Martin is forming strategic partnerships with both automotive and technology leaders. Access to advanced battery technologies, electric drivetrains, and software systems ensures that its vehicles remain competitive in an evolving landscape. By leveraging external expertise, Aston Martin can maintain its tradition of engineering excellence while embracing technologies that promise performance, reliability, and longevity in the electric era.

Part of this roadmap includes integrating autonomous features, connected vehicle technologies, and predictive analytics to enhance driving pleasure and safety. Aston Martin envisions a future where the driver's connection to the car is augmented by intelligent systems, providing seamless interaction without compromising the tactile thrill of driving. This approach ensures that even as vehicles evolve technologically, the **emotional, visceral, and aesthetic experience** of owning an Aston Martin remains uncompromised.

Strategic alliances have been instrumental in shaping Aston Martin's path forward. Partnerships with brands such as Daimler AG allow Aston Martin to integrate advanced infotainment, autonomous driving capabilities, and electrification technology directly into its vehicles. These collaborations ensure that Aston Martin maintains technological parity with other luxury manufacturers while preserving its unique identity.

Beyond traditional automakers, Aston Martin has been forging links with technology companies focused on artificial intelligence, machine learning, and sustainable energy solutions. These alliances allow the brand to experiment with **next-generation vehicle management systems, smart charging networks, and predictive maintenance algorithms**, positioning Aston Martin at the forefront of automotive innovation.

Strategic partnerships are also helping the company navigate global market expansion. By working with established players in emerging regions, Aston Martin can ensure that its vehicles are not only available worldwide but tailored to local tastes, regulations, and consumer expectations. From custom finishes to unique model editions, these collaborations enable Aston Martin to maintain both exclusivity and global relevance.

Aston Martin's future is inherently global. While Europe and North America remain core markets, there is enormous potential in Asia, the Middle East, and other emerging regions. These markets have seen a surge in demand for luxury automobiles, and Aston Martin is capitalizing on this trend by opening new dealerships, establishing flagship stores, and creating bespoke experiences for local clientele.

Expanding into global markets requires more than physical presence—it involves cultural understanding, localized design preferences, and service models that reflect the brand's luxury ethos. Whether offering **exclusive color palettes, interior**

customization options, or region-specific limited editions, Aston Martin is ensuring that each vehicle resonates with its owners while upholding the brand's heritage.

The company's expansion strategy is carefully balanced, seeking growth without diluting exclusivity. Maintaining limited production numbers and prioritizing quality over quantity ensures that each Aston Martin retains its sense of rarity and desirability. In doing so, the brand continues to cultivate an aura of aspiration and prestige worldwide.

For more than half a century, the V12 engine has been Aston Martin's hallmark—a symbol of unrelenting power, refinement, and sound engineering. Yet, as the industry shifts toward electrification, Aston Martin must reimagine its approach to powertrains.

Hybrid technology offers a bridge between the visceral thrill of combustion and the eco-consciousness of electric energy. Vehicles such as the **Aston Martin DBX SUV** and the **Valkyrie hypercar** illustrate how hybrid systems can amplify performance while reducing emissions. Looking ahead, Aston Martin plans to expand its hybrid offerings, with plug-in options and fully electric models gradually entering the portfolio. The brand's engineers are meticulously designing these systems to maintain the trademark driving experience—agile handling, powerful acceleration, and that unmistakable engine note—while integrating modern sustainability.

Fully electric hypercars and luxury sedans are also in development. By combining lightweight materials, advanced aerodynamics, and state-of-the-art battery technology, Aston Martin aims to ensure its electric vehicles achieve class-leading performance metrics without sacrificing luxury, comfort, or design. These next-generation vehicles will be testaments to the brand's enduring philosophy: **engineering meets artistry, speed meets sophistication**.

Aston Martin's hypercar program is central to its vision of the future. The **Valkyrie**, created in partnership with Red Bull Advanced Technologies, represents a milestone in automotive engineering—a vehicle where aerodynamic precision, hybrid propulsion, and advanced materials converge to redefine what a road car can achieve.

Future hypercars will continue this trajectory, integrating hybrid and electric powertrains to achieve unprecedented levels of acceleration, efficiency, and handling. These vehicles will serve as platforms for experimentation, testing technologies that may eventually trickle down into production models. Hypercars allow Aston Martin to continue pushing the envelope, capturing global attention and demonstrating that sustainability and performance are not mutually exclusive.

Each hypercar will be a showcase of technological mastery, blending AI-assisted performance optimization, advanced suspension systems, and lightweight composites. Yet, despite these innovations, the core driving experience—the connection between driver, machine, and road—will remain distinctly Aston Martin, ensuring that the brand retains its essence even in a radically evolving landscape.

While embracing innovation, Aston Martin's future remains deeply rooted in tradition. The company continues to prioritize **craftsmanship, aesthetic excellence, engineering precision, luxury, and heritage**. Every design decision, whether for an electric sedan or a hybrid supercar, is made with respect for the brand's legacy.

Aston Martin's cultural impact—bolstered by iconic appearances in James Bond films, music, and high-profile lifestyle associations—remains a guiding principle. The brand's vehicles are recognized not just as machines but as symbols of achievement, elegance, and sophistication. By maintaining this cultural resonance while introducing cutting-edge technology, Aston Martin ensures that its relevance transcends generations.

The next decades will see Aston Martin navigating a world where sustainability, performance, and luxury converge. Its roadmap includes expanding electric mobility, pioneering hybrid and hydrogen technologies, refining hypercars, and extending global market reach. Each step in this evolution is guided by a commitment to excellence, ensuring that every vehicle is unmistakably Aston Martin.

From fully electric Lagonda sedans to hybrid-powered supercars, Aston Martin is proving that a luxury brand can evolve responsibly without sacrificing its soul. The company is poised to continue capturing the imaginations of enthusiasts, collectors, and new generations of drivers worldwide.

Ultimately, Aston Martin is more than a car manufacturer. It is a testament to human ingenuity, a celebration of **luxury, artistry, technology, performance, and vision**. Its legacy is not just the sum of its cars but the cultural, technological, and emotional impact it has had on generations of enthusiasts.

As the brand steps confidently into the future, it will continue to define what is possible in automotive design and engineering. The journey ahead is filled with challenges—electric mobility, regulatory pressures, and global competition—but Aston Martin's commitment to innovation, excellence, and beauty ensures that it will thrive.

By 2050 and beyond, Aston Martin will remain at the forefront of automotive excellence, shaping the future while honoring its legendary past. Its cars will continue to inspire, captivate, and redefine the very meaning of performance and luxury. The brand's journey is far from over—it is merely entering a new and thrilling chapter in its illustrious history.

The automotive landscape of the future will not only be defined by electrification but also by **autonomous driving, connected systems, smart AI, and safety**. Aston Martin recognizes that the next generation of drivers will expect vehicles that are not only fast and luxurious but also intelligent, intuitive, and seamlessly integrated into daily life. Autonomous driving technologies, combined with real-time connectivity and artificial intelligence, are poised to redefine what it means to own and operate a luxury automobile.

Aston Martin is actively exploring adaptive AI systems that can learn driver preferences, optimize handling in real-time, and even predict potential road hazards before they occur. Imagine a car that understands the driver's style, adapts suspension settings mid-corner, and optimizes electric energy usage without intervention—this is the vision Aston Martin is working toward. Intelligent vehicle systems will also enhance safety, using predictive analytics and networked sensors to prevent collisions, improve navigation, and anticipate traffic conditions, all while maintaining the brand's emphasis on driver engagement and excitement.

While electric powertrains dominate the discussion about the automotive future, Aston Martin is also investigating

hydrogen fuel cells, biofuels, alternative energy, emissions reduction, and efficiency. Hydrogen presents a compelling opportunity for high-performance vehicles: zero emissions, rapid refueling, and high energy density suitable for hypercars or grand tourers.

Aston Martin engineers are experimenting with hybrid hydrogen-electric drivetrains that could combine the immediacy of electric torque with the long-range capability of hydrogen fuel. Early prototypes suggest that such vehicles could maintain Aston Martin's signature performance benchmarks while offering a truly sustainable powertrain. This research complements the brand's broader commitment to sustainability, ensuring that Aston Martin remains a leader in both high-performance engineering and environmental stewardship.

The interior of an Aston Martin has always been a sanctuary of **craftsmanship, luxury, comfort, design, and personalization**. Looking forward, interiors will evolve to reflect the changing priorities of drivers in the age of electrification and autonomy. Expect intuitive interfaces, augmented reality displays, and fully digital dashboards that blend seamlessly with handcrafted leather, polished metals, and exotic woods.

Customization will reach unprecedented levels, with clients able to choose unique color palettes, embroidered motifs, and materials that reflect individual tastes. Intelligent ambient lighting, adaptive seating, and immersive sound systems will create an environment where every drive is both a performance experience and a luxury escape. Aston Martin is defining the interior not merely as a cockpit but as a living space, harmonizing technology, comfort, and artistry.

Sustainability is no longer a peripheral concern—it is central to Aston Martin's future. Advanced materials such as **carbon fiber composites, recycled metals, lightweight alloys, bioplastics, and sustainable fabrics** are being integrated into vehicle design. These innovations not only reduce environmental impact but also improve performance by reducing weight and enhancing aerodynamics.

The company is also committed to carbon-neutral manufacturing, renewable energy in production facilities, and comprehensive recycling programs. Every Aston Martin produced in the future will reflect a marriage of eco-conscious engineering and uncompromised luxury. Sustainability is not a sacrifice—it is a philosophy embedded in every aspect of the car, from chassis to cabin.

Aston Martin is preparing for a world where cars are part of **digital ecosystems, integrated apps, mobility services, connectivity, and experience**. Owners will be able to manage their vehicles remotely, monitor performance, schedule maintenance, and even customize driving dynamics through advanced software interfaces. Smart connectivity will integrate with urban mobility solutions, offering route optimization, energy management, and predictive maintenance notifications.

Luxury, in this context, is not just about material opulence—it is about convenience, personalization, and seamless integration with a driver's lifestyle. Aston Martin's digital ecosystem will ensure that owning a car is as enriching as driving it, combining performance, sophistication, and intelligent connectivity in perfect harmony.

Even as Aston Martin embraces innovation, the brand remains committed to the enthusiasts and collectors who have sustained its legacy. Limited-edition models, bespoke commissions, and hypercar programs will continue to define the aspirational side of the marque. Collectors will have access to exclusive previews, factory tours, and interactive digital experiences, bridging the gap between classic craftsmanship and next-generation technology.

Aston Martin recognizes that its community is one of its greatest assets. As cars evolve, so too will events, clubs, and social experiences that bring owners together. From private track days to global luxury tours, Aston Martin ensures that its community remains connected, engaged, and immersed in the brand's ever-expanding universe.

The next decades will see Aston Martin solidify its position as a **global leader, innovator, luxury brand, sustainability, and performance**. Its vehicles will embody not only the pinnacle of automotive engineering but also the values of environmental responsibility, technological integration, and timeless design. By combining heritage with cutting-edge innovation, Aston Martin is crafting a roadmap that honors its past while embracing the future.

This vision includes not only electrification and hybrid technology but also the creation of vehicles that define new standards of luxury, connectivity, and experience. Aston Martin's ability to adapt, innovate, and inspire ensures that it will remain relevant, desirable, and iconic in a rapidly evolving world.

Ultimately, Aston Martin is more than a car company—it is a symbol of **innovation, luxury, heritage, performance,**

and aspiration. Its journey into the future is guided by a philosophy that blends respect for tradition with a commitment to cutting-edge technology and sustainability.

From electric hypercars to intelligent vehicles, from bespoke interiors to global mobility solutions, Aston Martin is shaping a future where performance, elegance, and responsibility coexist. The brand will continue to captivate enthusiasts, inspire cultural icons, and define the boundaries of automotive possibility.

As we look to 2050 and beyond, one thing is certain: Aston Martin will remain a benchmark of **luxury, engineering, artistry, excitement, and innovation**, a brand that continues to turn heads, inspire hearts, and lead the world into a new era of motoring excellence.

21: A LEGACY OF TIMELESS LUXURY

Aston Martin is more than a car; it is an institution. Since 1913, it has represented a rare combination of performance, sophistication, and exclusivity, earning the admiration of collectors, enthusiasts, and high-profile individuals worldwide. The name Aston Martin carries weight beyond the automotive industry—it evokes imagery of elegance, precision engineering, and a lifestyle steeped in refinement. For over a century, the marque has captivated hearts by fusing artistry with innovation, creating vehicles that are as much about the soul of design as the thrill of driving.

What distinguishes Aston Martin in the luxury automotive market is its unique ability to merge opulence with performance. While many manufacturers specialize in either extreme, Aston Martin consistently delivers cars that are meticulously crafted, exhilarating to drive, and breathtakingly beautiful. From the early AM models to the modern Vanquish, each Aston Martin embodies a careful balance of tradition and innovation, showcasing the finest materials alongside cutting-edge technology. This duality—the coexistence of heritage and modernity—is what allows the brand to remain timeless in a competitive landscape where trends come and go, yet Aston Martin endures.

The key to this longevity lies not just in engineering, but in creating an emotional connection with drivers. Owning an Aston Martin is a statement of taste, a declaration of achievement, and an entry into a community of like-minded aficionados who appreciate the blend of craftsmanship, luxury, and performance that defines the marque.

Every Aston Martin is a work of art. Its sleek lines, aerodynamic contours, and meticulously sculpted interiors reflect the brand's unwavering dedication to design excellence. From the DB5's debut in *Goldfinger* to the modern DB11, Aston Martin vehicles symbolize elegance and power, effortlessly conveying sophistication without ostentation.

The DB5, in particular, exemplifies this synthesis of design and performance. Introduced in the 1960s, it quickly became the quintessential luxury sports car, immortalized as James Bond's vehicle of choice. But the DB5 was more than a cinematic icon; it was a statement of automotive prowess, demonstrating Aston Martin's mastery of engineering and design. Today, the DB11, Vantage, and Vanquish continue this legacy, blending advanced technology with the timeless aesthetics that have defined Aston Martin for decades.

Performance is central to the Aston Martin experience. Each model offers exhilarating acceleration, precise handling, and a visceral connection to the road. Yet these characteristics are complemented by opulent interiors, hand-stitched leather, and bespoke finishes, ensuring that the thrill of driving is never divorced from the comfort and luxury that define the brand.

Driving an Aston Martin is an invitation into an exclusive lifestyle. It is a privilege reserved for those who value artistry, innovation, and heritage—a lifestyle that extends beyond the car itself into experiences, communities, and a legacy of excellence.

What truly sets Aston Martin apart is the meticulous craftsmanship that defines each vehicle. Cars are not simply assembled; they are constructed with a devotion to artistry that rivals fine watchmaking or haute couture. Every panel is hand-formed, every stitch in the interior hand-sewn, and every engine calibrated to perfection. This painstaking attention to detail ensures that each Aston Martin is unique, a reflection of the brand's ethos and the skilled artisans who bring it to life.

Craftsmanship at Aston Martin is not a marketing slogan; it is a philosophy. At the Gaydon headquarters in Warwickshire, highly trained artisans spend hundreds of hours perfecting every car, blending traditional skills with modern technology. From the polishing of aluminum panels to the creation of hand-finished dashboards, every vehicle carries the hallmark of dedication, patience, and artistry. This commitment ensures that even a century after its founding, Aston Martin continues to produce vehicles that stand as masterpieces of design and engineering.

Moreover, craftsmanship is embedded into Aston Martin's culture. Lionel Martin and Robert Bamford, the founders, instilled a philosophy of combining function with beauty. That philosophy remains unchanged today: technology and performance are never pursued at the expense of aesthetic elegance, and luxury is never divorced from engineering precision.

Ownership of an Aston Martin is more than possessing a car; it is joining an elite community. Through initiatives like Q by Aston Martin—where bespoke design choices are tailored to individual owners—the marque offers personalization that transcends conventional luxury. Owners can select everything from unique paint colors to custom interior materials, ensuring that every vehicle is a singular expression of its driver's taste.

The Aston Martin Owners Club, established decades ago, exemplifies the social dimension of the brand. Exclusive events, private track days, and driving tours create an immersive lifestyle, allowing owners to share their passion with like-minded individuals. Aston Martin owners are not simply clients; they are part of an extended family, ambassadors of a brand that values artistry, performance, and legacy.

This immersive experience extends beyond ownership. Attending an Aston Martin event is an education in luxury, design, and performance. Fans witness firsthand the meticulous process behind each car and interact with the engineers, designers, and craftsmen who ensure that every vehicle meets the brand's exacting standards. Aston Martin's commitment to experience fosters a connection that is both emotional and enduring, ensuring the brand's legacy is passed from generation to generation.

While steeped in tradition, Aston Martin has never shied away from innovation. The brand is embracing a future of sustainability without sacrificing the high-performance characteristics that define its identity. The development of electric and hybrid powertrains underlines Aston Martin's commitment to remaining at the forefront of automotive technology while honoring its heritage.

The revival of the Lagonda brand marks a pivotal moment in Aston Martin's evolution. Lagonda, now a fully electric luxury marque, embodies the company's vision for sustainable performance. By introducing zero-emission vehicles without compromising on elegance or performance, Aston Martin demonstrates that luxury and responsibility can coexist. From hybrid powertrains in the DBX SUV to the anticipated all-electric models, innovation is seamlessly integrated into the brand's DNA.

At the same time, performance remains uncompromised. Every new model continues to push the boundaries of engineering, incorporating advanced materials, aerodynamic designs, and state-of-the-art technology. Aston Martin's vehicles are not only beautiful—they are thrilling to drive, reflecting a balance of tradition, artistry, and cutting-edge innovation.

Aston Martin's legacy is not merely historical; it is a living, evolving testament to excellence. While the marque honors its past, it continuously reinvents itself to meet the expectations of contemporary luxury consumers. Every model, from the DB11 to the Valkyrie, reflects a delicate balance of tradition and innovation. The company's engineering teams work tirelessly to ensure that each car is not only a marvel of performance but also an embodiment of timeless design.

Craftsmanship remains central to Aston Martin's identity. From the hand-finished leather interiors to the bespoke paintwork, every detail is scrutinized and perfected. Yet innovation is never sacrificed for artistry. By integrating modern technologies, Aston Martin ensures its vehicles deliver cutting-edge performance, dynamic handling, and future-ready features that appeal to a global clientele. This dual commitment—honoring heritage while embracing technological advancement—is a hallmark of the brand and the reason its appeal remains undiminished over a century later.

Beyond engineering excellence, Aston Martin has cultivated a unique global community of enthusiasts. Owning one is not simply about having a luxury car—it is about joining a worldwide family that celebrates craftsmanship, innovation, and shared passion.

From the Aston Martin Owners Club (AMOC) founded in 1935 to local chapters in North America, Europe, and Asia, the network of clubs connects owners across continents. These clubs host exclusive events, driving tours, and social gatherings, providing an environment where passion for the marque is shared, nurtured, and celebrated.

Each club chapter emphasizes camaraderie and education. Members often gain privileged access to Aston Martin factories, design studios, and testing tracks. They witness firsthand how each car is crafted, how engines are fine-tuned, and how every aesthetic detail is perfected. The clubs not only foster a sense of belonging but also ensure that the brand's values are passed down through generations, maintaining a living legacy of Aston Martin excellence.

No discussion of Aston Martin's cultural impact is complete without its iconic association with cinema. The DB5 in *Goldfinger* was not just a prop—it became a symbol of style, sophistication, and refined power. This cinematic connection elevated Aston Martin from a luxury car manufacturer to a global cultural icon.

The brand's appearances span decades, from Bond films like *Casino Royale* and *Skyfall* to blockbusters where the vehicles symbolize wealth, precision, and elegance. Musicians, artists, and fashion designers also reference Aston Martin as a marker of success and elite taste. Whether in music videos, lyrics, or red-carpet appearances, the marque is synonymous with high-end living.

This cultural footprint amplifies the emotional appeal of Aston Martin. Ownership is thus not just about driving a high-performance car; it is about participating in a lifestyle, a shared recognition of prestige, artistry, and cultural significance.

The future of Aston Martin is increasingly tied to innovation, particularly in sustainability, electrification, and advanced engineering.

Hybrid technology has already been integrated into models such as the DBX707 and Valkyrie, providing a blend of combustion and electric power for enhanced efficiency without compromising performance. These hybrid systems allow drivers to enjoy electric power in city driving, while retaining the exhilaration of traditional engines on the open road.

Meanwhile, the Lagonda brand represents Aston Martin's commitment to fully electric luxury vehicles. With the Vision Concept, Lagonda offers a glimpse into a world where performance, design, and sustainability converge. Every curve, every detail, and every technological component is designed to maintain Aston Martin's signature elegance while addressing the environmental challenges of the 21st century.

Aston Martin continues to redefine the limits of automotive performance through hypercars. The Valkyrie, co-developed with Red Bull Racing, exemplifies this ambition. It combines hybrid technology, lightweight materials, and

Formula 1-inspired aerodynamics to produce an unparalleled driving experience. Future hypercars will further integrate electrification, advanced battery systems, and artificial intelligence for driving optimization, ensuring that the brand remains at the cutting edge of automotive innovation.

Hypercars are also experimental platforms, allowing Aston Martin to test new technologies for subsequent production models. Innovations in materials, aerodynamics, and electronics eventually trickle down into consumer vehicles, ensuring every model benefits from state-of-the-art advancements. This approach guarantees that performance, craftsmanship, and exclusivity are never compromised, even as the brand adapts to a new era of sustainable motoring.

Aston Martin's future is also shaped by its global strategy. Beyond Europe and North America, emerging markets in Asia and the Middle East present significant growth opportunities. The brand is expanding its dealership network, offering bespoke vehicles tailored to regional preferences, including unique finishes, exclusive editions, and custom interiors.

Strategic partnerships are equally crucial. Collaborations with Daimler AG provide access to advanced infotainment, autonomous driving systems, and electrification technology. Partnerships with tech companies allow Aston Martin to explore innovations in AI, machine learning, and next-generation vehicle connectivity. These alliances ensure the marque remains competitive, forward-looking, and capable of integrating modern technological advancements into its luxury offerings.

While performance and luxury remain non-negotiable, Aston Martin recognizes the global imperative of sustainability. Hybrid and electric powertrains, lightweight materials, and carbon-neutral manufacturing are central to the brand's roadmap.

The Gaydon production facility is embracing renewable energy sources, advanced recycling, and sustainable manufacturing processes, ensuring the environmental footprint is minimized. These efforts reflect a broader vision: creating high-performance vehicles that are environmentally responsible, without diminishing the signature Aston Martin driving experience. Sustainability and luxury, once thought at odds, now coexist seamlessly under the Aston Martin ethos.

Looking ahead, Aston Martin continues to blend tradition with innovation. Its vehicles will increasingly feature electrification, advanced driver-assistance systems, and hypercar-inspired performance. Yet every model will remain unmistakably Aston Martin: elegant, powerful, and exclusive.

The marque's philosophy extends beyond technology—it encompasses culture, lifestyle, and experience. Whether through ownership clubs, exclusive events, or cultural presence in films and music, Aston Martin maintains a human connection, ensuring that its legacy is not only mechanical but also deeply emotional.

In the end, Aston Martin is more than a car manufacturer—it is a symbol of aspiration, craftsmanship, and achievement. Its century-long history demonstrates resilience, innovation, and an unwavering commitment to excellence. The future promises electrification, hypercars, and global expansion, but the core values—luxury, performance, and elegance—remain constant.

From bespoke interiors to cutting-edge powertrains, from cinematic appearances to global communities of enthusiasts, Aston Martin continues to define what it means to create timeless luxury. The brand's legacy is secure: a testament to human ingenuity, artistic vision, and engineering mastery. As it drives into the future, Aston Martin remains an enduring icon, a benchmark of excellence, and a celebration of what is possible when passion meets innovation.

22: ASTON MARTIN RACING

Aston Martin's reputation extends far beyond the realm of luxury and elegance. Its very essence is entwined with **the pursuit of speed**, precision, and engineering mastery. While its road cars evoke awe and desire, its racing pedigree proves that the brand was engineered to **win under pressure**. Since its earliest days, Aston Martin has embraced motorsport as both a proving ground for its technology and a stage to demonstrate resilience, innovation, and the brand's indomitable spirit. From endurance classics like Le Mans to modern FIA World Endurance Championship campaigns, Aston Martin Racing has become synonymous with excellence and the relentless drive to succeed.

The passion for racing is embedded in the DNA of the marque. Every chassis, engine, and aerodynamic design tested on the track informs the refinement of road-going Aston Martins. In this way, competition is not merely a sport—it is a laboratory for performance, innovation, and the uncompromising pursuit of perfection.

Long before Aston Martin was a household name, co-founder Lionel Martin envisioned a brand that could **conquer hills, circuits, and races**. From the company's inception in 1913, racing was central to Aston Martin's identity. Lionel Martin himself was a competitive driver, and his early cars were designed not just for style but for endurance, handling, and precision on the road.

By the 1920s, Aston Martin was competing in hill climbs and small-scale endurance races across Europe. The 1922 French Grand Prix marked a crucial step onto the international stage, placing the brand among the competitive elite. Through the 1930s and 1940s, Aston Martin cars repeatedly demonstrated their engineering superiority, competing in events that tested reliability, speed, and driver skill. These formative years established the philosophy that would carry the marque through decades: that **luxury need not compromise speed**, and elegance could coexist with raw performance.

The pinnacle of Aston Martin's early racing success came at the **24 Hours of Le Mans** in 1959. Piloted by Carroll Shelby and Roy Salvadori, the Aston Martin DBR1 achieved one of the most celebrated victories in motorsport history. Competing against powerhouse manufacturers like Ferrari and Porsche, the DBR1 demonstrated not just speed but durability, strategy, and flawless engineering.

Ted Cutting's design genius was evident in every aspect of the car. Lightweight construction, a powerful inline-six engine, and meticulously calculated aerodynamics combined to create a vehicle that was perfectly balanced for endurance racing. Victory at Le Mans cemented Aston Martin's racing legacy, proving that the marque could dominate on one of the most demanding stages in automotive sport. That triumph has since become a touchstone for the brand, referenced in marketing, storytelling, and the DNA of every modern Aston Martin.

Following decades of intermittent motorsport activity, Aston Martin returned to professional racing in **2004**, partnering with Prodrive to establish Aston Martin Racing (AMR). This strategic initiative signaled the brand's renewed commitment to endurance racing and GT competition, blending historic expertise with modern innovation.

The first car to emerge from this renewed vision was the DBR9, a GT1-class racer derived from the road-going DB9. Its debut at the **2005 Sebring 12 Hours** marked the beginning of a new era, one that would see Aston Martin reclaim prominence on tracks worldwide. Victories followed, most notably at Le Mans in the GT1 category in 2007 and 2008, demonstrating that Aston Martin was once again capable of producing championship-winning cars.

The DBR9 laid the foundation for AMR's dominance in endurance racing, combining **raw power with refined aerodynamics**. It was soon followed by the Vantage GTE, a car purpose-built for FIA World Endurance Championship (WEC) competition. The Vantage GTE became Aston Martin's flagship in endurance racing, racking up multiple class wins at Le Mans in 2014, 2017, and 2020, as well as contributing to Manufacturer's Championship victories in 2016 and 2020.

The evolution of these cars underscores a philosophy: every technological advance on the track translates into **enhanced performance for road vehicles**. From lightweight construction to adaptive aerodynamics, lessons learned in GT racing influence models such as the Vantage and DBX, bridging the gap between the track and the open road.

Aston Martin's relationship with Formula 1 is historic yet cyclical. Initial forays in 1959 and 1960 were short-lived, hindered by an underpowered engine. Fast forward to 2021, and Aston Martin returned to the world's premier motorsport category with the Aston Martin Cognizant Formula One Team, now the **Aston Martin Aramco F1 Team**.

Led by Lawrence Stroll, the team invested in state-of-the-art facilities, technical expertise, and elite driving talent. Names like Sebastian Vettel, Fernando Alonso, and Lance Stroll brought credibility and competitiveness to the grid. The partnership with Mercedes-AMG for power units and technology gave the team a cutting-edge advantage. Early podium finishes reinforced Aston Martin's commitment to engineering excellence and high-performance racing, proving that the brand's motorsport ambitions were alive and thriving at the highest level.

Aston Martin Racing has always served as a crucible for innovation. The racetrack is where **concepts become realities**, and technologies developed for competition often transition seamlessly to road cars.

Key innovations include:

- **Lightweight carbon fiber chassis**, which first optimized performance in endurance racing and later enhanced road models like the One-77 and Valkyrie.
- **Advanced aerodynamics**, derived from FIA WEC and F1 insights, improving downforce, cornering, and stability.
- **Hybrid and electric powertrain research**, ensuring Aston Martin stays ahead in the era of sustainable high-performance vehicles.

This commitment to technological advancement ensures that every Aston Martin benefits from the lessons learned under the extreme conditions of motorsport.

Aston Martin Racing today represents a perfect marriage of **heritage, engineering, and vision**. The brand continues to compete at the forefront of endurance, GT, and Formula 1 racing, remaining a formidable presence across multiple disciplines. The philosophy remains unchanged: every race car is an expression of the company's identity, a combination of cutting-edge technology, strategic ingenuity, and a relentless desire to win.

From the roar of engines at Le Mans to the precision of Silverstone, Aston Martin Racing demonstrates that performance is inseparable from tradition. Every victory, podium finish, and technological breakthrough reinforces the brand's position as a leading innovator in motorsport.

Looking ahead, Aston Martin is embracing **hypercar competition** with models like the Valkyrie AMR Pro. These vehicles merge Formula 1 technologies with endurance racing principles, creating cars that are faster, lighter, and more efficient than ever before.

The hypercar program is more than just a showcase of speed; it is a testbed for the future. Developments in aerodynamics, powertrains, and materials provide insights that will trickle down to road-going models, ensuring that Aston Martin continues to **blend luxury with advanced engineering.**

Aston Martin is also responding to global calls for sustainability. From hybrid endurance cars to the exploration of synthetic fuels, AMR is investing in eco-conscious technologies that reduce environmental impact without compromising performance.

The goal is clear: to **lead motorsport into a sustainable era**. By integrating green technologies, Aston Martin is positioning itself to remain competitive while aligning with the broader automotive industry's shift toward responsibility and innovation.

GT racing remains a cornerstone of Aston Martin's strategy. The brand's commitment to FIA WEC and other global endurance competitions ensures that **its presence on the track is unmatched**. With consistent class wins, championship pursuits, and ongoing development of race-ready Vantage models, Aston Martin Racing continues to uphold the tradition of excellence that defines its DNA.

Every race is both a demonstration of skill and a laboratory for technological advancement. Drivers, engineers, and designers collaborate in an unbroken chain of innovation, ensuring that lessons learned on the circuit improve the performance, safety, and efficiency of Aston Martin's road vehicles.

Ultimately, Aston Martin Racing is not merely a division—it is **the heartbeat of the brand**. Its victories, innovations, and relentless pursuit of excellence encapsulate what it means to be an Aston Martin. From historic Le Mans triumphs to contemporary Formula 1 podiums, racing defines the brand's spirit, reinforcing its reputation as a creator of vehicles that are both luxurious and extraordinarily capable.

The track remains a proving ground for the **engineering brilliance, craftsmanship, and vision** that make Aston Martin one of the most revered names in automotive history.

Every victory adds a new chapter, every innovation ensures the brand remains relevant, and every race car carries forward a legacy of uncompromising excellence.

For Aston Martin, motorsport is inseparable from identity. Racing shapes design, inspires innovation, and strengthens the connection between brand and enthusiast. As Aston Martin moves into the future—with electric, hybrid, and hypercar technologies—the same philosophy endures: the pursuit of victory, excellence, and timeless craftsmanship.

Whether on the endurance circuits of Europe, the streets of global Grand Prix venues, or the next generation of hypercar tracks, **Aston Martin Racing embodies ambition and legacy**, ensuring that the spirit of competition remains at the very heart of the brand for decades to come.

Aston Martin Racing is not just about **winning on the track**; it is a laboratory where automotive innovation thrives. Every race car, from the DBR1 to the Valkyrie AMR Pro, embodies cutting-edge engineering designed to push the boundaries of performance. Aerodynamics, lightweight construction, advanced suspension, and hybrid powertrain technologies are all developed and tested under the harshest conditions before ever reaching a road car.

The symbiotic relationship between racing and road vehicles ensures that **every technological breakthrough** on the track enhances the luxury, speed, and driving experience of Aston Martin's production cars. For instance, carbon fiber monocoques and advanced energy recovery systems pioneered in endurance and hypercar programs have directly influenced models like the Valkyrie and the DBS Superleggera. This continuous cycle of innovation keeps Aston Martin at the forefront of performance, blending artistry with engineering precision.

The Valkyrie AMR Pro exemplifies Aston Martin's commitment to **hypercar performance and innovation**. Built with Formula 1-inspired aerodynamics, hybrid-electric powertrains, and extreme lightweight materials, it represents the pinnacle of racing technology applied to the road. These hypercars serve a dual purpose: they compete at the limits of performance in racing categories while acting as technical flagships, showing what is possible for future road-going Aston Martins.

Beyond speed, the Valkyrie AMR Pro demonstrates how **sustainability and performance coexist**. Aston Martin is exploring synthetic fuels, hybrid drivetrains, and efficient energy management systems, ensuring that hypercar racing evolves responsibly. By merging environmental consciousness with cutting-edge performance, Aston Martin positions itself for decades of relevance in both motorsport and luxury automotive markets.

Aston Martin Racing's influence is truly global. Its strategy extends beyond European endurance circuits to include participation in high-profile GT championships worldwide. By competing in markets across North America, Asia, and the Middle East, AMR not only enhances brand recognition but also strengthens Aston Martin's connection with international automotive enthusiasts and collectors.

This global vision is paired with bespoke engineering, tailored race cars, and strategic partnerships, creating vehicles **optimized for every track and climate**. Collaborations with Prodrive, Mercedes-AMG, and other technical partners ensure access to the latest performance technology, giving Aston Martin a competitive edge in every racing arena it enters.

The next decade will see Aston Martin Racing leading in **green technology and motorsport sustainability**. Hybrid endurance cars, energy-efficient engineering, and the development of alternative fuels ensure that the brand remains competitive while addressing environmental challenges. These initiatives demonstrate that Aston Martin Racing is not just looking backward at its storied history but also forward toward a cleaner, faster, and more responsible future.

Sustainability is now embedded into the engineering DNA of Aston Martin Racing. Every component—from energy recovery systems to hybrid powertrains—is tested not only for performance but for efficiency and environmental impact. This forward-thinking approach guarantees that the next generation of Aston Martin race cars will embody **speed, style, and responsibility simultaneously**.

Aston Martin's racing achievements have influenced more than technology—they have shaped **brand perception and global identity**. Success on the track enhances the allure of road cars, creating a powerful narrative: owning an Aston Martin means participating in a legacy of speed, precision, and triumph.

From the historic DBR1 Le Mans victory to contemporary Formula 1 podiums, racing stories infuse Aston Martin with mythic qualities that transcend automobiles. The brand is not only a symbol of luxury but also a testament to human ambition, engineering mastery, and competitive excellence. Through racing, Aston Martin connects emotionally with drivers, collectors, and enthusiasts around the world.

Looking ahead, Aston Martin Racing is preparing for an era of **hypercar, electric, and hybrid dominance**. Key initiatives include:

- **Valkyrie AMR Pro Expansion:** Pushing hypercar performance and aerodynamic innovation even further.
- **Hybrid and Electric Development:** Integrating cutting-edge technologies from F1 and endurance racing to future road models.
- **Sustainable Fuel Programs:** Aligning with global environmental targets while maintaining competitive edge.
- **Global GT Dominance:** Continuing to win class championships and FIA WEC titles worldwide.

These projects reflect Aston Martin's philosophy: innovation, craftsmanship, and performance are inseparable. By blending tradition with futuristic technologies, the brand ensures that racing excellence continues to define its identity.

Aston Martin Racing represents the **heartbeat of a legendary brand**. From early hill climbs to Le Mans triumphs, from GT dominance to Formula 1 returns, racing has shaped the company's identity in profound ways. The racetrack is not just a venue—it is a proving ground where every detail of performance, engineering, and design is tested under the most extreme conditions.

The DNA of competition informs every road car, ensuring that elegance and speed coexist seamlessly. Every victory, podium finish, and technological innovation feeds back into

Aston Martin's heritage, reinforcing the brand's promise of luxury, exclusivity, and unmatched performance.

As Aston Martin looks to the future, racing remains integral to its mission. **Hypercar innovation, electric performance, and sustainability** will drive the next era of competition. The brand's global reach, technical partnerships, and visionary engineering ensure that Aston Martin will remain a force in both motorsport and luxury automotive markets for decades.

The story of Aston Martin Racing is a story of passion, vision, and excellence. Every car on the track is a statement of intent, every innovation a testament to the brand's relentless pursuit of perfection. Racing is not just a chapter in Aston Martin's history—it is the lifeblood that ensures the marque continues to inspire, captivate, and dominate in the world of high-performance automotive engineering.

Ultimately, Aston Martin Racing embodies the values that make the brand extraordinary: elegance, performance, and **uncompromising pursuit of perfection**. Its past achievements, present innovations, and future ambitions demonstrate that motorsport is more than a sport—it is the very foundation of Aston Martin's identity.

For enthusiasts, collectors, and drivers alike, Aston Martin Racing represents the ultimate combination of luxury, speed, and technical mastery. As long as cars roar across circuits, engines scream, and aerodynamics push the limits of possibility, **Aston Martin will always lead**, inspiring generations to chase excellence on the track and the open road.

www.ingramcontent.com/pod-product-compliance
Lightning Source LLC
LaVergne TN
LVHW052305200926
848481LV00002B/471
9798232162498